S0-AZU-338

STORM OVER RANGELANDS

STORM OVER RANGELANDS

PRIVATE RIGHTS IN FEDERAL LANDS

BY
WAYNE HAGE

A Project of the
Free Enterprise Legal Defense Fund

FREE ENTERPRISE PRESS

BELLEVUE

STORM OVER RANGELANDS

THIS BOOK IS COPYRIGHT © 1989, 1990, 1994 BY WAYNE HAGE. ALL RIGHTS RESERVED. NO PART OF THIS BOOK MAY BE REPRODUCED IN ANY FORM OR BY ANY ELECTRONIC OR MECHANICAL MEANS IN-CLUDING INFORMATION STORAGE AND RETRIEVAL SYSTEMS WITHOUT WRITTEN PERMISSION, EXCEPT IN THE CASE OF BRIEF QUOTATIONS EMBODIED IN CRITICAL ARTICLES AND REVIEWS.

THIRD EDITION
Published by the Free Enterprise Press

Typeset in Times Roman on AMSi computers by the Free Enterprise Press, Bellevue, Washington. Front cover photo by Patrick Morey, courtesy Barkley's Art Company. Book design by Northwoods Studio.

The Free Enterprise Press is a division of the Center for the Defense of Free Enterprise, 12500 N. E. Tenth Place, Bellevue, Washington, 98005. Telephone 206-455-5038; FAX 206-451-3959. The opinions expressed in this book are not necessarily those of the Center, its board of directors, or its members.

This book distributed by Merril Press, P. O. Box 1682, Bellevue, Washington, 98009. Telephone 206-454-7009. Additional copies of this quality paperback book may be ordered from Merril Press at $14.95 each.

LIBRARY OF CONGRESS CATALOGING-IN-PUBLICATION DATA

Hage, Wayne, 1936-
    Storm over rangelands : private rights in federal lands / by Wayne
Hage. — 3rd ed.
        p.   cm.
    "A project of the Free Enterprise Legal Defense Fund."
    Includes bibliographical references and index.
    ISBN 0-939571-15-3 : $14.95
    1. Pasture, Right of—West (U.S.) 2. Grazing districts—West
(U.S.) 3. Range management—Law and legislation—West (U.S.)
4. Range policy—West (U.S.) I. Title.
KF5630.H34   1994
346.7304'674—dc20
[347.3064674]   94-8516
                                                                          CIP

PRINTED IN THE UNITED STATES OF AMERICA

CONTENTS

Acknowledgments

This book is a project of the Free Enterprise Legal Defense Fund, which provided major funding. The Fund is a division of the Center for the Defense of Free Enterprise, a non-profit foundation. The Fund was established to defend the free enterprise system through the courts.

The Fund operates three programs:

Education: Studies and analyzes government actions for possible legal challenge, and publishes the results in books and conferences.

Legislation: Provides policy analysis for consideration of legislative bodies on methods of protecting free enterprise, private property rights, and the wise use of resources.

Litigation: The Fund files "Friend of the Court" briefs on behalf of individuals who are harassed by big government and opponents of free enterprise. The Fund also refers individuals to competent legal counsel, and initiates test cases against those who would violate the right of Americans to participate in the free market.

Storm Over Rangelands was funded in part by generous grants from the following:

Mr. and Mrs. Frank Duran
Helendale, California

Bert and Amelia Smith
Ruby Valley, Nevada

Joe and Sue Fallini
Tonopah, Nevada

Key Agricultural Credit Corporation
North Salt Lake, Utah

Nevada Agricultural Foundation
Reno, Nevada

Harry and Joy Wilson
Wilson Ranch, Inc., Denio, Nevada

Von and Marian Sorensen
Wells, Nevada

N-6 Grazing Board
Battle Mountain, Nevada

Steve and Carol Wilmans
King City, California

Eugene B. Casey Foundation
Gaithersburg, Maryland

Editor's Introduction to the Third Edition

It is a great pleasure to be able to write an introduction to the Third Edition of this book as proof of its durability. The demand for Wayne Hage's *Storm Over Rangelands* has been strong and steady. The critical commentary from experts across the country has been astute and most helpful in correcting and rearranging the text to make it more useful. Most of the changes in this edition go beyond the improvements in the Second Edition, and encompass additions suggested by attornies interested in solidifying the basic thesis of private rights in federal lands. Readers will find many significant changes throughout the work.

As I wrote about the first and second editions, *Storm Over Rangelands* is a true original. Author Wayne Hage unfolds here a completely new picture of the American federal lands and the western rancher's place on those lands. Some will perhaps characterize Mr. Hage as a revisionist. That would be a predictable but unjust assessment, for he has not so much reinterpreted known history as he has pieced together for the first time an unknown history that has been laying unpublished on the scattered shelves of dusty archives for more than a century. Wayne Hage's story of the federal lands is very different from earlier historiography because he refused to ignore what other historians have refused to see.

Wayne Hage is a successful western stockman who holds an advanced degree in biological science and is well versed in history. His family and that of his wife Jean have been in the West since Gold Rush days. Long exposure to family recollections and anecdotes provided Mr. Hage with a keen sense of history. Constant contact with academics as a trustee of the University of Nevada Foundation and in many other academic associations cultivated in Mr. Hage intellectual rigor in many disciplines. Although he is not a professional historian, I found his manuscript of *Storm Over Rangelands* to be not only sound in method but also path-breaking in approach.

Mr. Hage's book is so much a breakthrough in conservation history that I submitted the manuscript in sections to highly qualified impartial academics for editorial review and criticism. Dr. Alfred Runte, formerly of the University of Washington faculty and now writing and publishing independently, reviewed the first three chapters and critiqued Mr. Hage's methodology and hypotheses. Dr. Harold K. Steen, executive director of the Forest History Society and Professor of History at Duke University, reviewed the chapter on forest reserves and kindly gave permission to quote materials from the *Journal of Forest History*. Dr. Lawrence Rakestraw, distinguished historian of federal lands issues, reviewed the entire manuscript.

Editorial review of the legal theses was provided by Harry Swainston, Assistant Attorney General, State of Nevada; John Marvel of Marvel and Hansen, Elko, Nevada; Robin Rivett and James Burling of Pacific Legal Foundation, Sacramento, California; Karen Budd of Budd-Falen, Laramie,

Wyoming; William F. Schroeder, Vale, Oregon; and by attorneys Eric Twelker and Marti Albright.

The criticisms of all these experts have been thoroughly considered by Mr. Hage and are greatly appreciated. However, any errors of fact or judgment in the finished product should in no way reflect upon the reviewers.

Storm Over Rangelands is based upon the critical examination of archival sources—many published here for the first time—and published sources, the selection of particulars from authentic materials, and their synthesis into a narrative intended to stand the test of critical methods.

Because *Storm Over Rangelands* breaks so much new ground, Mr. Hage and I have endeavored to insert documentation or oral corroboration from authoritative sources for every major premise and every important fact in the manuscript. Nor has Mr. Hage uncritically repeated his sources. Mr. Hage and I have tried to judiciously assess the capabilities of other workers in the field of federal land history in relying upon their publications. Mr. Hage and I agreed that he would not accept on trust any facts, and certainly not any current interpretation, of published historians of the federal lands.

Because *Storm Over Rangelands* is a true original, even Mr. Hage's conclusions have undergone similarly rigorous testing. While I have left untouched Mr. Hage's suggested solutions to federal lands problems, I have insisted that all conclusions be supported by persuasive documentation or, where remoteness in time and intentional secrecy by the historical actors have prevented the recovery of documents, strongly linked sequences of circumstantial evidence. Mr. Hage has included a Table of Cases at the end of the text that will assist those interested to locate the numerous legal citations used throughout the book. Despite our attempts at rigor, we are all too aware of the many shortcomings in the text. We must leave the final assessment of this book's success to our readers and future researchers.

Yet a tentative prognosis is not inappropriate. Like any original thesis, Wayne Hage's *Storm Over Rangelands* will undoubtedly raise more questions than it answers. Future historical researchers will find in it a treasure trove of leads and clues that can stimulate whole careers. This book's admirers will likely read more into it than it says, but that is the fate of authors. Its detractors will likely mischaracterize its contents and meaning, but that too is the fate of authors. The one thing nobody interested in the federal lands issue will likely do is to ignore it.

I wrote in earlier editions, "After *Storm Over Rangelands* has swept away a century of ignorance and misinformation about our federal lands—and let the future show whether I exaggerate—America will never be the same again." The U.S. Forest Service took this book seriously enough to systematically harass the author, precipitating one of the most important lawsuits on property rights in American history: *Hage v. United States,* still in court. The government then trumped up a felony conviction against the author, which was overturned by the Ninth Circuit Court of Appeals in 1994. They can't destroy the message of this book so they try to destroy the messenger. They're not having much luck at that, either.

Ron Arnold
Bellevue, Washington

Author's Preface

I have long been concerned with the incessant attack on the western range livestock industry. The assault begun by the conservation movement in the 1890s continues today under key environmentalist leaders and the government agencies they politically control. During my long effort to understand this endless barrage of anti-grazing rhetoric, I discovered that the attack on western livestock interests was not a mindless manifestation of urban ignorance but rather a well-planned, superbly orchestrated attempt to destroy the property rights of western interests. The hue and cry over protection of natural resources and preservation of pristine wilderness has been the stalking horse behind which environmentalists pursue control of the West's vast resource riches with the help and financing of large eastern corporations. The question never was one of preservation of natural resources as opposed to development of those resources. The basic question from the beginning of the conservation movement up to today is, "Who is going to develop those resources?"

In most historical accounts of the western federal lands I found no one asking fundamental questions about how the original intent to build America into a land of small private holdings resulted in one-third of our nation ending up as a permanent capital asset of the federal government. Nor did I find anyone asking who would benefit from such pervasive government ownership or why corporate interests have funded nature preservationists so generously for more than a century. Simply stated, most histories do not recognize the enigma posed by vast federal ownership of the West's land base at all, but unquestioningly accept the premises that the conservation movement of the late 19th century was responsible for stopping the progressive privatization of the public domain in a two-part effort, one led by well-meaning citizens to preserve a pristine environment and the other by technocrats to build a "gospel of efficiency" into America's resource policy.

A few adventurous historians such as Alfred Runte have proposed original answers to the riddle of public lands. Runte formulated the "useless lands" thesis to explain the existence of national parks, but his work did not extend to the much larger area of rangeland and forest reserves and cannot explain their existence. Others such as David A. Clary have suggested that social movements such as the populists had a significant effect on resource policy, and some recent writers of "the new labor history" such as Daniel Cornford and Ian Radforth are asking interesting questions about corporate capitalists' role in resource policy. There is a vast literature that describes *how* the federal lands came to be retained by the government, but there is no convincing explanation of *why*. And there is a great deal of fantasy even about the *how*.

When I began to dig beyond the myth, factual information about the real objectives of the preservation movement was abundant—and shocking. I have written what I found. No doubt some will take issue with the historical and legal theses of this book. If no one disagreed with arguments such as mine there would be little further research and consequently little further progress in this vital field that so demands extensive exposure. Critics must now deal with this question: Are there private rights in federal lands? I would like to believe that critics will be rigorous, not ignore that question, and base their critique on *what* is right rather than *who* is right.

This book could not have been written without the assistance of many people. On the top of the list is Ron Arnold, editor in chief of the Free Enterprise Press. Ron pursued may of the leads I provided, fleshing out little-known historical vignettes with extensive documentation. The research that identified the "cattle capitalists" is largely Ron's. The historical discussion of the crucially important forest reserve act is largely the result of Mr. Arnold's research with the able help of his daughter Andrea.

I must also thank my own daughter Ramona for research in the Library of Congress and the National Archives in Washington, D. C. Much of the primary source material she found is published here for the first time.

My thanks also to Janet Arnold, Ruth Agee, Margaret Gabbard, and Sabrina Agee for their hours spent checking facts and correcting errors.

Those who reviewed the manuscript, Dr. Alfred Runte, Dr. Harold K. Steen, Dr. Lawrence Rakestraw, Harry Swainston, John Marvel, Robin Rivett, James Burling, Karen Budd, William F. Schroeder, Eric Twelker, and Marti Albright all have my deepest gratitude. Constance Brooks of Linsey, Hart, Neil & Weigler originally provided stimulus for much of my legal discussion.

Steven Hanke of Johns Hopkins University, Baltimore, Maryland, was the intellectual leader of the privatization movement of the early 1980s and gave me many useful ideas. Charles Cushman, executive director of the National Inholders Association, must be credited for my original contact with Alan Gottlieb, president of the Center for the Defense of Free Enterprise, who agreed to accept my manuscript for publication by the Center's division, the Free Enterprise Press.

The Nye County Law Library was an invaluable source of information. I am grateful to William P. Beko, District Judge, Fifth Judicial District, State of Nevada, for making its resources available to me. Important research assistance was provided by Philip Dunleavy, John Adams, and Jeff Morrison of the Nye County District Attorney's Office.

My deepest appreciation to the hundreds of supporters who subscribed for copies of this book to help make its publication possible. This book is yours in more ways than one.

Last, but certainly not least, a special thank you must go to my wife Jean for the endless hours she spent on the word processor as we went through several drafts of this book over a period of nearly three years.

Whatever value this book may have must be credited to these fine people. Any errors of fact or judgment must be attributed to me alone.

Wayne Hage
Tonopah, Nevada

PART ONE

ROOTS OF CONFLICT

1

A STORM GATHERS

Range war! Here and now! No, it's not the Old West. It's not the clash of cattlemen against sheep herders or ranchers against sodbusters in time warp. It's today's headlines about Congress threatening to increase stiff grazing fees ranchers already pay on federal land in the eleven western states. It's Forest Service and Bureau of Land Management officials sharply curtailing grazing permits to broaden their regulatory powers. It's Public Broadcasting System television specials about environmentalists accusing ranchers of overgrazing, and pressing to eliminate the livestock industry from federal lands with slogans such as "Livestock-Free By '93!"

It's this and a thousand battles more that the news doesn't cover, all boding devastation for America's livestock industry—and for every other resource extraction industry in the nation, from minerals and timber to fisheries and petroleum. Thus the fallout from this storm over rangelands is confined neither to the range nor to the West. The political and economic destiny of all Americans is unalterably hitched to this multi-faceted strife.

Today's range war is for control of our nation's greatest storehouse of natural resource wealth—the federal lands. Shall that control be centralized in the hands of federal officials or decentralized to give markets, users and the general public a say in the land's allocation? The outlines of this conflict have been etched in the public's mind for more than a decade as a familiar "developers versus environmentalists" clash, or a "Sagebrush Rebellion" asserting "states' rights over federal tyranny." This book is not about either characterization.

The thesis of this book is that valid private property rights exist in federal lands, a finding with profound implications. A previously ignored cloud on the title of the West's federal range lands may well stir an even bigger conflict over land and resources than any in our nation's history. At the center of this developing strife stands a single, simple, unasked question: Who owns the range?

THE UNASKED QUESTION

The view from Table Mountain's 11,000-foot crown is incomparable. The immense mountain panorama evokes feelings beyond words. The space stretches your imagination. I don't recall which one of my children first called it the "edge of the world," but as you ride up to the precipice where the eastern verge drops away into Little Fish Lake Valley some four thousand feet below, you certainly get that feeling. And all you can see out there is ranch.

1

Our family owns the range rights here on Table Mountain. We also own the water rights. The land itself is classified as federal land. The national government owns it. The Forest Service, a branch of the Department of Agriculture, administers most of what's visible from here. The Bureau of Land Management, a division of the Department of the Interior, administers other land where we own the range rights. Two other large agencies also administer federal rangelands, the Fish and Wildlife Service and the National Park Service, both under the Interior Department. Grazing on Fish and Wildlife Service lands is permitted in places, but under stringent restrictions. Livestock grazing is almost always forbidden in national parks. Smaller but significant federal areas, including some rangelands, are administered by the Department of Defense, the Nuclear Regulatory Commission (the old Atomic Energy Commission) and the Bureau of Reclamation. That's the tangle of bureaucracy that surrounds my private range rights and water rights.

Thus the range before us is part of that vast federal domain which so dominates the eleven states west of the 100th meridian today. Of course, not all lands in the western states are federal lands. Much of the West is privately owned the same as in the East. More than 50 percent of the land in California is privately owned—55.4 percent, to be exact. Even though less than half of Oregon is privately owned (52.2 percent federal ownership), more than two-thirds of Washington is in private hands (28.8 percent federal land). Some states such as Nevada, Idaho and Utah, though, have a very high percentage of federal lands. Nevada is 86 percent federally owned, Idaho 63.1 percent, and Utah 64.1 percent. But throughout the eleven western states, federal ownership of *resource* lands—those areas valuable for commodity production—is a dominant and dominating fact of life.[1]

Table Mountain and the world you can see from it lies in a region that some say history forgot: It was left out of early America's exuberant drive to settle the public domain east to west and place it in private hands. No land settlement law passed by Congress ever really fit it—not the Preemption Acts, the Homestead Act, the Timber Culture Act, the Timber and Stone Act, or the various Desert Land Acts—none of them.

In a way you could say it was leftover land. It was mostly mountains and desert and high plains: too dry for farming. It was certainly useless under the impractical terms of the 1862 Homestead Act (free land for settlers given away in 160 acre parcels, which were not big enough to cultivate much of anything within regions of scant rainfall). No one really wanted it very badly in the early days of America's westward expansion. As we shall see in succeeding chapters, during the course of history settlers and commodity producers decided they did want it and Congress passed a number of laws to convey various rights for the use of the federal domain to private individuals. In 1866 Congress recognized western water rights previously established under local law and custom on the federal lands. Numerous state laws and court cases during the 1880s and '90s extended those rights to cover nearby land. Laws passed in 1897 and 1934 created a livestock permit system that recognized private rights to graze specified federal lands. That is how I ended up with the range rights and water rights. I pay a federal tax on the grazing rights. It's called a grazing fee.

This high country is excellent summer range. If a cow is left alone, she can make a good living on these mountain meadows and bring a fat calf off with her in the fall. Deer do well here also. There is a lot of natural cover and browse. Massive stands of quaking aspen and endless thickets of mahogany sumach, pinion pine, and juniper make a home for deer, sage grouse, bobcats, mountain lions, and a host of other wildlife. If you like brook trout, almost every stream up here has a nice population of "natives." Generally, the federal government has regulatory control of migratory wildlife and the states have regulatory control of resident wildlife.[2]

From the edge of the world I can look back to the west where Freeport Minerals Company is exploring for gold in Antone Canyon. I own the range and water rights there too, but Freeport owns the mineral rights. A little further to the west is Round Mountain, one of the world's largest open pit gold mines. It produced an all-time record of 59,052 ounces of gold in the second quarter of 1988 alone and totaled 190,578 ounces in 1987.[3] Most of this production is on federal lands, but Echo Bay Mines is the legal owner of Round Mountain's mineral rights. They employ a regular workforce of more than 500 men and women at Round Mountain, generating net earnings of $35.2 million in 1987, many millions of which went directly into the local economy. Echo Bay recently donated a school to the local community. They have always been good neighbors.

Even farther to the west, hidden behind the high crest of California's Sierra Nevada, loggers harvest national forest timberlands and foresters perpetually renew them. Timber operators large and small own contracts to purchase and remove federal timber.

In the opposite direction, far to the east I can see the tops of the Quinn-Grant range. In Railroad Valley—just this side of the ridge—lie the largest producing onshore oil wells in the continental United States. The wells themselves are on federal lands, but rights to the oil are owned privately by petroleum companies.

Beyond the southeastern horizon over the Papoose Range and near Las Vegas lies Lake Mead National Recreation Area. The National Park Service has regulatory control over the ground, but private concessioners own the buildings, hotels, service stations, campgrounds, and other facilities they construct for public use.

THE SPLIT ESTATE

From up here on the edge of the world we can begin to see that the ownership of these "federal" lands is not at all the seamless fabric of government title we might have imagined. The high estate of ownership in the federal lands is split between a number of parties, both private and governmental, which embrace a number of rights including water rights, grazing rights, mineral rights, wildlife rights, petroleum exploration rights, and timber harvest rights.

Each of these rights is secured by some statute or court decision. My water rights are guaranteed by the 1866 law I mentioned above, as well as by a Nevada statute that is similar to other state water laws throughout the West.[4] My grazing right is secured by the Taylor Grazing Act of 1934, which was shaped by many state laws that recognized the prescriptive right of the first grazer to establish an "accustomed range" good against all newcomers, and by the Organic Act of 1897.[5] The mining companies' deed of title to subsurface rights came from the Mining Act of 1872. Timber firms have contract rights to harvest trees on federal lands based upon nearly a hundred years of law culminating in the National Forest Management Act of 1976. The Concession Policy Act of 1965 gives national park concessioners a "possessory interest" in the buildings, hotels, service stations, campgrounds, and other facilities they construct for public use.[6] These are all protected private property rights on federal lands.

Today, federal agencies such as the Bureau of Land Management and the Forest Service argue that ranchers have no "property rights" in the federal lands, only "grazing privileges" conveyed by a permit which can be revoked. It should be clear that the point is arguable from the facts recited above, but, for the sake of further demonstration, consider four additional features of today's split estate rangelands:

1. If a joint-private/federal rancher decides to sell, the federal range which his private ranch controls is validly transferred by the same deed that transfers the private land and improvements.

2. Range rights on the federal range can only be purchased from the rancher who owns them, not from the federal government.

3. If the military takes a rancher's federal range for military use, then military administrators are required by law to enter civil condemnation proceedings in open court and pay for the rancher's interest in the federal range.

4. When a rancher dies, the Internal Revenue Service is entitled to levy an estate tax on the rancher's ownership interest in the federal lands.

Given these four facts, certain conclusions are inescapable. If federal lands are routinely transferred to a new owner by a private deed of sale, a *private property right* of some sort in the federal lands is obviously involved. If you can only buy range rights from a private rancher and not from the government, a *private property right* of some sort in the federal lands is obviously involved. If the military has to pay a private rancher condemnation awards for his interest in federal lands, a *private property right* of some sort in the federal lands is obviously involved. If the IRS can tax a rancher's ownership interest in the federal lands, a *private property right* of some sort in the federal lands is obviously involved. And when you repeat the fact that you can't buy range rights from the federal government, the conclusion is clear: The federal government owns the federal lands, but it does not own range rights to the federal lands. *The private rancher owns the range rights to the federal lands.*

COMMENSURABILITY

And then there's the completely private land. The private land we can see from up here on the edge of the world generally follows the meadows along the creeks and streams in the valleys and lower ranges. Most of these lands were privatized under one homestead law or another not long after the Civil War.

These private parcels form the nucleus of the large ranches that operate in the area. Most of the living quarters, livestock handling facilities, holding fields for range livestock, and supplemental feed production such as hay and grain for the range livestock, are concentrated on these small pieces of private ground in the valleys and lower ranges. Irrigation ditches and their rights of way to convey water to these lands were withdrawn from the public domain and recognized by various Acts of Congress as private property. Rights of way for roads and highways giving access to these lands were also withdrawn by the same Acts. Some of those private improvements are located on the federal lands, but generally speaking, it is these small parcels of more productive private lands that make the surrounding federal rangelands usable.

If it were not for these commensurable pieces of developed private property being used for homes and winter livestock feeding, the federal rangelands would have little value. Conversely, if it were not for the federal rangelands where nutritious grasses and forbs provide a necessary grazing complement, the small pieces of private land would have a lesser value. By owning a small well-developed area of private "base land," a rancher may acquire permanent and exclusive legal grazing rights to commensurable federal lands. This principle of allotting range rights according to ownership of commensurable base land was recognized by the federal government in the first regulations issued by the Forest Service after its creation in 1905, and formalized for other public domain lands in the Taylor Grazing Act of 1934.[7]

The complementary relationship between small developed private holdings and vast federal grazing lands emerged when the first settlers came to the high plains of the West in the mid-1800s and subsequently shaped the law of range rights. Rights to grazing land were crucial. But water was the key to everything.

WATER

While the pioneer immigrants of the early 1850s pushed across Texas and the Kansas and Nebraska and Dakota Territories, they pushed west of an unseen boundary that spelled doom for their farming way of life. As historian Robert A. Caro explained this invisible line in the Texas Hill Country:

> The line was an "isohyet" (from the Greek: *isos*, equal; *hyetos*, rain)— a line drawn on a map so that all points along it have equal rainfall. This particular isohyet showed the westernmost limits in the United States along which the annual rainfall averages thirty inches; and a rainfall of thirty inches, when combined with two other factors—rate of evaporation (very high in the Hill Country), and seasonal distribution of rainfall (very uneven in the Hill Country, since most of it comes in spring or autumn thundershowers)—is the bare minimum needed to grow crops successfully. Even this amount of rainfall, "especially with its irregular seasonal distribution," is, as the United States Department of Agriculture would later state, "too low" for that purpose. East of that line, in other words, farmers could prosper; west of it, they couldn't. And when, in the twentieth century, meteorologists began charting isohyets, they would draw the crucial thirty-inch isohyet along the 98th meridian. . . .[8]

runs down the United States, north to south, at the 98th meridian. You could not farm by normal methods "west of 98, west of thirty inches of rain." As western historian Walter Prescott Webb summed it up, "When people first crossed this line, they did not immediately realize the imperceptible change that had taken place in their environment, nor, more is the tragedy, did they foresee the full consequences which that change was to bring in their own characters and in their modes of life."[9]

What became clear in the 1840s, though, was that there were valuable mineral deposits in the West. Men found gold in California in '49, they crossed the Sierra Nevada to the Comstock Lode in pursuit of silver in '59, and they rushed to Butte searching for copper in '62. Miners had spread throughout the mountain West by the 1870s, in Montana, Colorado, Utah, New Mexico, Arizona.

The western mines had a direct influence on the development of the livestock industry. Somebody had to feed all those miners. More often than not it was some Forty-niner or Fifty-niner who discovered that mining could be extremely hard work, and raising livestock on the grass near the boom towns was more suited to his talents.

The advancing railroads also had a direct influence on the western livestock industry. Some of the immigrant farmers from the fertile East who crossed 98 and watched their crops wither in the summer dust read the message correctly and decided that grazing cattle to feed the trackside merchant towns was more suited to the soil's talents. Those who fed the miners and those who fed the railside towns soon realized that all the dwellers of the Eastern Seaboard ate beef and wore shoeleather, as did the British export market. There were definite possibilities hiding in the grass west of 98, west of thirty inches of rain.

Yes, the grass. Grass enough to give a perceptive Westerner visions of a boundless future. In 1868, Nathan Addison Baker, editor of the Cheyenne *Leader*, predicted:

> That a future of the greatest importance is in store for the western plains, no one who has traveled over and lived upon them for any considerable length of time can doubt ... this country will make stock raising a specialty, since it is one of the most profitable branches of industry which it is possible for civilized man to engage in.[10]

Perhaps Mr. Baker overestimated the riches to be won in stock raising, but he perfectly expressed the view of his day. It was a view of hope, and hope is what drew men to the early West, hope for gold and silver and copper and grass. All those things were there waiting for them, but when they got there, they discovered that the one vital ingredient missing was a law suited to the land. The law was most deficient in the area of water.

INFORMAL PROPERTY ARRANGEMENTS

The process by which our pioneers settled west of thirty inches of rain fell outside any established legal framework. Congress had granted farmers homesteads that might have been suitable east of 98 but were just too small west of 98. West of 98, west of thirty inches of rain, there was not enough water.

Congress had created the General Land Office in 1812 to deal with disposing of the public domain into private hands. In 1849 the Department of the Interior was established to cope with the enormous responsibilities involved with the public lands. A poor man could enter the public domain in Ohio, Illinois or Missouri and legally take up for himself a 40 to 160 acre farm, depending on how generous Congress was feeling in its latest land disposal law. Administering the disposal of the vast public domain sometimes stacked up paperwork several years behind actual events. "Doing a land office business" became a catch phrase for any runaway work load. Millions of farmers got a livable place from the government's great storehouse of public domain land.

Congress did not confer similar thoughtful largesse upon the miner and the rancher. When those thousands of miners in the mid-1800s came to Sutter's California from around the world searching for gold, there was no land disposal law for them. They became squatters on the federal lands, outside the established legal order, with no government to impose order or to settle their disputes.

These pioneers found themselves faced with Rousseau's classical social contract situation of being in "a state of nature" in which "the war of all against all" threatened to drown the promise of golden wealth in a pandemonium of gunfights, personal vendettas, and vigilante retribution. They literally had to make rules or die.

The first rule they all settled upon in the gold streams was the familiar "first come, first served." It not only established private rights over what had previously been held in common—specifically, minerals and water—but it also established limitations on what any single individual could own.[11]

Legal scholar Charles W. McCurdy explained:

> Following a tradition of collective action on the mining frontiers of other continents, the miners formed districts, embracing from one to several of the existing 'camps' or 'diggings' and promulgated regulations of marking and recording claims. The miners universally adopted the priority principle, which simply recognized the superior claims of the first arrival. But the . . . miner's codes defined the maximum size of claims, set limits on the number of claims a single individual might work, and established regulations designating certain actions—long absence, lack of diligence, and the like—as equivalent to the forfeiture of rights. A similar body of district rules regulates the use of water flowing in the public domain.[12]

The unanimity of scholarly opinion on how western water rights evolved is revealed by comparison with another authority's statement of essentially the same facts. Nationally respected water rights expert Wells A. Hutchins tells the story of California, where western water law first appeared:

> Gold was discovered in the foothills of the Sierra Nevada, California,

in January 1848. This development and the resulting mining industry had a profound influence upon the political and economic growth of California and on the development of water law throughout the West. As water was required in much of the gold mining processes, rights to the use of the water were of fundamental importance. This mineral area was Mexican territory when gold was discovered but was ceded to the United States less than 6 months later by the Treaty of Guadalupe Hidalgo. There was no organized government there in the early years, nor much law except that made by the miners who helped themselves to the land, gold, and water under rules and regulations of their own making as they went along. In the words of the United States Supreme Court, speaking through Justice Field who had been Chief Justice of California, the miners "were emphatically the law-makers, as respects mining, upon the public lands in the State."[13]

The rules and regulations of the miners were made by and for the individual camps and hence varied from one locality to another, but essentially the principles that they embodied were of marked uniformity. These principles related to the acquisition, holding and forfeiture of individual mining claims, based upon priority of discovery and diligence in working them. And to the acquisition and exercise of rights to the needed water were applied comparable principles—posting and recording notice of intention to divert a specific quantity of water, actual diversion and application of water to beneficial use with reasonable diligence, continued exercise of the right, priority in time of initiating the appropriation, and forfeiture of priority for non-compliance with the rules—in other words, the doctrine of prior appropriation of water for beneficial use. These property rights in land and water were thus had, held, and enjoyed under local rules and were enforced by community action.[14]

Hutchins concludes that today's "long detailed 'water codes,' with their centralized administrative procedures, developed inevitably from these early brief declarations of a few basic principles." These "few basic principles," customs actually, were adopted by miners, irrigators, and cattlemen as they took possession of the most valuable portions of the federal lands without legal authorization from the federal government. The rules were simple and well known: The first person to work a mine or divert water from a stream for stock water and file a proper claim acquired a prior right to what was taken. The right to take water was based upon making "beneficial use" of the resource taken—you couldn't merely appropriate something to hoard or keep away from others or speculate upon. If you failed to make beneficial use of your claim, you forfeited your rights to it. The customs became universal. Later settlers respected them.

Such informal property arrangements are not unusual historical developments. They are commonly brought forth by the disorderly circumstances of pioneering. Germanic mining law from the Middle Ages was developed from a similar "mining freedom."[15] The process was typical around the world in the 19th Century—Australia, Africa, America. First the pioneers made their own rules man-to-man. Then local organizations were convened to elaborate and enforce the rules, as was the case with innumerable miners' and stockmens' organizations in the early American West. When disputes arose over ownership of mineral claims or water rights, as they inevitably did, they tended to be settled not in flashy gunfights, but rather in prosaic frontier courts. When the territory was admitted to statehood, its new legislature gradually formalized these informal property arrangements into statutes that became the basis of later state court decisions. Thus the process that began as a mad scramble in the 1840s and '50s to prove who owned what in the goldfields finally gained recognition and protection by the power of state law.

The federal government soon recognized these informal property arrangements in an Act of Congress: At the initiative of Nevada Senator William Stewart, Congress passed the Act of July 26, 1866, giving consent to the state laws and local customs on which private claims to minerals and water rested. It had this to say specifically about water:

> ... That whenever, by priority of possession, rights to the use of water for mining, agricultural, manufacturing, or other purposes, have vested and accrued, and the same are recognized and acknowledged by the local customs, laws, and the decisions of courts, the possessors and owners of such vested rights shall be maintained and protected in the same . . .[16]

THE PRIOR APPROPRIATION DOCTRINE

The principle of carving for oneself a private domain out of the public domain is known as the "prior appropriation doctrine," or "arid region doctrine," and was recognized not only by Congress, but also by state constitutions, legislation, and judicial rulings as the West was settled. For example, in 1851 the legislature of California, in one of its first official acts, sanctioned the local customs by which water and mineral rights had been established.[17] A series of California court decisions gave additional sanction to the prior appropriation doctrine.[18] Nevada courts handed down a similar series of rulings, as did many other states.[19]

Among the most important cases was an 1878 U.S. Supreme Court decision in which the court held that the object of the 1866 federal statute for protection of private water rights "is to give the sanction of the United States to possessory rights which had previously rested solely upon local customs, laws, and decisions, and to prevent such rights from being lost on sale of the lands."[20] This ruling effectively conveyed clear title in prior appropriation water rights to private parties and their heirs and assigns forever.

The prior appropriation doctrine became firmly established in the arid West by 1875, being adopted by state or territorial statute, or recognized by high court decision, or both, in Arizona, California, Colorado, Montana, Nevada, New Mexico, and Wyoming. By 1900 the prior appropriation doctrine had been authorized in Idaho, Kansas, Nebraska, North Dakota, Oklahoma, Oregon, South Dakota, Texas, Utah, and Washington. The prior appropriation doctrine in those states displaced partially or totally the old common law riparian doctrine—the right of everyone through whose land a natural watercourse runs to the benefit of the stream as it passes through his land. The Nevada Supreme Court was so vehement in its support of the appropriation doctrine that it wrote in a 1902 decision, "The doctrine of riparian rights is so unsuited to conditions which exist in the State of Nevada, and is so repugnant in its operation to the doctrine of appropriation, that it does not prevail in Nevada."[21]

Riparian law essentially tied one's right to water to an existing right in land through which a natural watercourse ran. The new prior appropriation doctrine, "a revolution in the law of water,"[22] effectively separated rights to water from rights to land: A prior appropriator could move water from a streambed for great distances in order to use it for irrigating farms, for stock watering ponds, or for domestic use. Acts passed by Congress in 1866, 1891 and 1901 allowed for the withdrawal of ditches, flumes and pipelines and their rights of way from the public domain as private property. Ironically, the prior appropriation doctrine led to the tying of existing rights in water to rights in the land—without water the land would be worthless—the exact reverse of the riparian doctrine.

As William D. Rowley wrote in his *U.S. Forest Service Grazing and Rangelands: A History:*

> Those graziers who had arrived earlier attempted to establish prescriptive rights to the range by their customary occupancy and acceptance of those rights among themselves, their stock organizations, and roundup committees. Some western writers have compared these range rules to the customs of the mining districts in their attempt to regulate and distribute access to ore in rich stream beds or quartz lodes. As with mining-district rules, western legislatures sometimes reinforced the regulations of stock organizations by authorizing inspections of cattle shipments for legitimate brands and passing water laws that gave rights to the first users. It was in the area of water rights that early arrivals established their most effective power. Whoever controlled the water sources often controlled great areas of land.[23]

CUSTOMARY RANGE RIGHTS

In other words, if a settler properly appropriated water on some part of the public lands, that prior water right formed the basis for claiming prescriptive rights to the nearby range. The appropriation of water on federal lands gave claimants control of grazing lands within the distance livestock would normally travel to and from water. It was a cow country axiom that water controlled the range. Cattle were able to travel many miles to water, but they had to find it somewhere, and he who had a prior appropriated water right had an effective claim on the adjacent range.

There is no such thing as range without water. As the Nevada Supreme Court said in a 1936 range-dispute case, "We have a right to take judicial notice of matters of public knowledge, such as the climatic and range conditions in this state, and that livestock cannot exist upon the public range unless there is water available for them to drink."[24]

Once a settler had a properly appropriated water right, then tenure or "customary occupancy" of the nearby range (i.e., putting it to beneficial use by grazing his livestock on it over time) could confirm his claim to exclusive use of that range in a court as against any later claimant.

Historical documents bear witness to the fact, as in this sworn statement of a New Mexico rancher:

> A custom has grown up and become thoroughly established among people of this community that where one stockman has developed water on and taken possession of the range by fully stocking the same that he will not be molested by other stockmen in his possession and enjoyment of such range.[25]

H. M. Taylor, head of the Bureau of Animal Industry, commented in his 1886 annual report:

> It will be seen that the ownership of the watering places gives tenure to contiguous range. This fact is recognized by Western cattlemen, and the question as to the number of cattle individual owners are permitted to hold, under regulations of the various local associations, is determined by the question of water frontage.[26]

As the western miners had enforced their early rules by convening local associations, so did the ranchers. Livestock associations developed in every western state, and by the early 1880s, there were even several rival national associations vying for influence in the highly fragmented and sectionalized industry. The ranchers knew they must look to themselves for a workable plan to protect the range and their rights in it.

A detailed rationale for local livestock association regulations was given by historian Ernest Staples Osgood:

> The presence of others along a stream, too long to permit individual control, meant that the exclusion of outsiders must come through some sort of an understanding among those already on the ground. Cooperation among neighbors in the conduct of their business resulted in the growth of a certain amount of range privilege and good will. Participation in the roundup, in the use of the common corrals, in the group protection against Indians, thieves, and predatory animals and, in some cases, in the group drive of the beef turnoff to the railroad could be permitted or denied to the newcomer. To deny such privileges, often appeared to be the only way of preventing the overcrowding of a range already taxed to its full carrying capacity. The success of such a method would, of course, depend upon the size of the outfit so denied and upon the amount of cooperation among the older stock growers.[27]

Thus the simple principles of prior appropriation water rights, customary occupancy of the range, and cooperation with your fellow stockmen were adopted as the primary rules. Local association rules in many cases were adopted by state legislatures as range law. The associations themselves thereby gained a quasi-governmental status among ranchers. Wyoming even passed an 1884 law that vested all range privileges and good will in the Wyoming Stock Growers' Association.[28]

But in 1903 water authority Elwood Mead looked back to the 1880s and described the fly in the ointment:

> There was no law by which men could legally secure control of the land they occupied. All the land laws dealt with farming land. There was no provision for leasing or settling the grazing land in tracts large enough to be of any service. Hence the range stockmen simply took possession of the country. Each man chose a location which suited him, fixed in a rough way the boundaries of his domain, and helped to create a public sentiment which made it unpleasant, if not dangerous, for a late comer to attempt to share with him the territory he had so marked out. In this way range rights came to have the force of law.[29]

That force of law, however, was still tenuous, based as it was on the 1866 Act that recognized state laws, court cases and local customs, but without explicit federal law giving title to settlers for ranch-sized areas. Stockmen relied upon informal property arrangements, i.e., they commanded grazing land without owning title to it, because there was no legal way they could obtain title to it. Homestead laws had been drawn up by eastern legislators making law on the erroneous assumption that eastern conditions extended into the West. But failure to understand the arid climate west of 98 was not the only reason Congress did not make suitable law. As historian Richard Bensel points out, "sectional competition—grounded in a geographical division of labor between the economically advanced northern core and the underdeveloped southern and western periphery—has been and remains the dominant influence on the American political system."[30]

Many eastern legislators during the booming westward expansion of the 1880s did not want to see the West become politically and economically powerful, for strictly sectional reasons. Among the most blatant statements of this widespread feeling was New Jersey Congressman William McAdoo's retort to western colleagues during an 1888 floor debate on a bill to ease access to public lands: "I think our western friends are making a very great mistake. This country is developing rapidly enough. There is no difficulty in the United States about disposing of the surplus lands. The natural increase of our population is tremendous, and the phenomenal immigration which is pouring upon our shores makes it, in my opinion, the right policy that we should rather stop and retard than advance the settlement of the public lands."[31] Such sentiments were normally cloaked in more sanctimonious rhetoric; McAdoo's statement is remarkable for its clear anti-western and anti-immigration bias. This deep current of sectionalism underlies all the debate on public lands, then and now.

The chief land law problem for grazers was the limited acreage any one person could legally obtain from the public domain—Homestead acts invariably denied acquisition of acreages large enough to accommodate livestock grazing. Various "desert land acts" in the 1870s allowed arid-land settlers to claim 640 acres of land instead of the usual homesteader's 160 acres, but this was hopelessly inadequate for most ranching purposes, which might require several thousand acres to support an economic herd.[32] Despite continuous lobbying efforts, the cattlemen never managed to convince Congress to provide a method whereby the rancher, like the miner and the farmer, could obtain absolute title to viable lands for his occupation. Even though the ranchers had followed the normal pattern of customary property arrangements that the farmers had pursued to the Homestead Act of 1862 and the miners had pursued to the Mining Law of 1872, there was never a Ranching Law of eighteen-hundred-and-anything.

Nonetheless, the need for a federal range rights law for stockmen was widely recognized. It was advocated in 1878 by no less a figure than John Wesley Powell, the one-armed Civil War veteran famed for his daring 1869 exploration of the Grand Canyon of the Colorado, and as director of the U.S. Geological Survey from 1881 to 1894. In his landmark study *Report on the Lands of the Arid Region of the United States*, Powell recommended that Congress allow viable ranch units of 2,560 acres, or four miles square. Powell also recognized the central importance of water, and recommended that standard survey grid lines be abandoned in arid land so that property lines could follow watershed lines, or some other demarcation more in tune with the nature of the country. He also suggested irrigation works, the protection of watersheds, and equitable systems for water distribution to be devised through community action.[33] His proposal was never enacted into law.

Presidents Grant and Hayes also noted the need for land use legislation to accommodate the demands of the arid West. Their proposals were never enacted into law.

As the years passed, the livestockman, pursuing the same principles of preempting the public domain as the miner and farmer, began to realize he had failed in his efforts to acquire fee title to the land. It is important to understand why he failed, and what the consequences were—and are.

SEEDS OF CONFLICT

Four major events during the latter half of the nineteenth century shaped the federal rangelands problem in the western United States today.

1. The American Civil War and the profound impact it had in intensifying sectional politics.

2. The emergence of an extensive class of wealthy entrepreneurs in the victorious northeastern quadrant of the country.

3. The financial Panic of 1873 followed by the economic depression of the 1880s.

4. The emergence of an obscure movement known as conservationists.

While hundreds of thousands of Union and Confederate soldiers were dying on eastern battlefields, far western settlement was quietly progressing. The far western population of the 1860s included many emigrants from the southern states who had come with the California gold rush a decade earlier. Prior to the discovery of gold in California, most United States gold mining had taken place in the Southeast, primarily Georgia. When gold was discovered at Sutter's mill it immediately attracted experienced southeastern gold miners. A disproportionate number of the Forty-niners came from South Carolina, Georgia, and Alabama. These miners combined with other southern settlers who moved west for different reasons to comprise a significant population segment with solid southern sympathies on the nation's western shore.

Some northerners, like many southerners, moved west when they saw conflict coming in the period before the Civil War. Some migrated to escape the conflagration after it had started. Not all northerners were solid Lincoln men. Not all were against slavery. Many wanted a peaceful accommodation with the South, not war. These anti-Lincoln "Copperheads," as they were called, came under considerable political pressure in the Northeast as the war progressed. Civil rights were suspended in the North for a period during 1861 and 1862. People could be arrested and jailed without being charged with a crime. Many were. Many of them were Copperheads. Some Copperheads fled to Canada or other foreign countries. Many migrated to the West.

By the end of the Civil War the victorious northeastern quadrant of the nation ruled supreme over a defeated and devastated South. The northeasterners controlled the political and financial powers of the nation. The West needed both financial help and political support from the northern core, a West that was peopled to a large degree by northern Copperheads and southerners. There was a tendency in the victorious northeastern political circles to view the West's population as politically unreliable if not actually disloyal.

It was in this climate that the nation's new financiers began to dominate western development. The West needed their money and the northern core financial interests were willing to provide money in return for control of the West's resource base. The South had its carpetbaggers, the West had its plutocrats. Dozens of eastern capitalists poured money into western ventures. Much of it went into the livestock industry.

Livestock interests representing the northern core capitalists found themselves late-comers in the race for prior appropriation of land and water in competition with the earlier settlers. In order to disenfranchise the earlier settlers and usurp prior appropriation rights for themselves, these cattle capitalists sometimes intentionally overgrazed the range. Many original settlers found their range destroyed by large herds of cattle and sheep owned by some northeastern interest. Often their only choice was to sell what was left to the eastern competitors—including stockwater rights on the federal lands and associated prior appropriation claims.[34]

The real objective of the northeastern financiers was control of the land base and all its resources including minerals, timber, and hydroelectric sites. Ownership of the prior appropriation claims to stockwater and range rights appeared to be the surest path to eventual ownership of the land itself.

The Panic of 1873 led to a full-scale depression during the 1880s. Northeastern capitalists could now pick and choose among the West's assets as ranchers, farmers, and other settlers went bankrupt on a wholesale basis. The depression was, of course, nationwide, but even that played in favor of the northeastern capitalists. The harsh economic hardships suffered by the populace east of the Mississippi were conveniently blamed on "cheap production" from the West. The public believed it and the eastern capitalists plundered the West with little political interference from a Congress and courts dominated by eastern solons and judges.

In an effort to mask this wholesale rape of the West with a "public interest" facade, eastern capitalists poured money into the fledgling nature preservation branch of the conservation movement. Railroad magnates such as Edward H. Harriman doled out largesse—including extensive expense-paid expeditions—to nature writers such as John Burroughs and John Muir, the latter of whom unwittingly helped Harriman obtain a monopoly of the railroad route to Yosemite National Park. Conservation-minded publisher Robert Underwood Johnson of *Century* magazine helped set the pubic agenda for nature preservation—not realizing he was being used as a catspaw by the very men he courted for support: George Vanderbilt, Clarence King, Professor Alexander Agassiz, Theodore Roosevelt, and many other eastern capitalists who owned western cattle and mining operations.[35]

The most obvious result of this capitalist-conservationist coupling was the creation in 1891 of the forest reserves that evolved into our present vast system of national forests. Once the eastern capitalists realized in the late 1880s that prior appropriation rights had blocked their efforts to monopolize the West, they supported the idea of forest reserves. To the capitalists, forest reserves were a way to keep competitors from controlling the resources while they found political means to secure that control to themselves. To the nature conservationists, forest reserves were an expression of their aesthetic-religious fervor for natural preservation. From about 1889, numerous eastern capitalists supported the nature conservationists because nature preservation provided a respectable mask for their hidden objectives.

BUFORD v. HOUTZ

An obscure quarrel between two stockmen out West intensified the capitalists' interest in forest reserves. In Utah Territory the owners of Promontory Stock Ranch Company, M. B. Buford, J. W. Taylor, and the two Crockers, Charles and George, were having a problem in the late 1880s. Two sheepherders, John S. Houtz and Edward Conant, along with others, grazed their herds totalling about 200,000 sheep over the open public domain range that lay intermixed in alternate sections with Promontory's private land, where Mr. Buford and his company grazed some 20,000 head of cattle. Since any two public domain sections meet only at an infinitesimally small point, sheep inevitably set hoof on Promontory's property, not to mention the huge roving surges of sheep that wandered far afield to graze Promontory's private forage.

Mr. Buford sued Mr. Houtz in the Third Judicial District Court of Utah Territory in and for Salt Lake County.[36] Mr. Buford and his partners argued that because Mr. Houtz did not own any of the lands involved and because he could not prevent his sheep from straying from the public domain onto Mr. Buford's private cattle range that the sheepherders should be enjoined from the use of such public lands for pasturage. Mr. Buford complained that he could not build fences around his checkerboarded private land without thereby unintentionally (and unlawfully) enclosing the lands of the United States and also cutting off his own cattle from necessary water sources.

Mr. Houtz, on the other hand, alleged that he had a perfect right to use the public domain range because there is an implied license, growing out of the custom of nearly a hundred years, that the public lands of the United States, especially those in which the native grasses are adapted to the growth and fattening of domestic animals, shall be free to the people who seek to use them, when left open and uninclosed, and no act of government forbids this use.

Further, Mr. Houtz argued that the government of the United States had known of this use, had never forbidden it, nor taken any steps to arrest it, but had consented to and encouraged it. And besides, in the United States there had never been a law saying a man was bound to keep his cattle confined within his own grounds or else he would be liable for their trespasses upon the uninclosed grounds of his neighbors.

The U.S. Supreme Court on February 3, 1890, held that there was no equity in allowing plaintiffs Buford et al. to deprive defendants Houtz et al. of this recognized right to permit their livestock to run at large over the unoccupied lands of the United States and feed upon the grasses found in them, and under pretense of owning a small proportion of the land in controversy, obtain for themselves the monopoly of this valuable privilege. Buford lost.

The court also agreed with Mr. Houtz that there is an implied license to use the federal lands for grazing. That set a precedent of profound importance which has not been overturned to this day. It forged a vital link in the chain of rights that ranchers hoped would one day give them direct title to the federal range. It said they had an implied license to use the land. Finally the federal government, through the Supreme Court, had recognized the right of ranchers to use the federal lands. Now the rancher had two avenues of approach to obtaining full title to the range. First, the federal water law of 1866 recognizing informal property arrangements. Second, the case of *Buford* v. *Houtz* asserting an implied license to use the federal lands for grazing purposes. Another step or two, and ranchers would have their equivalent of the Homestead Act and the Mining Act.

But it was not to be. The hope for full title in the range was all but extinguished the very next year by a law nobody understood.

FOREST RESERVES

Ironically, what is today called the "Forest Reserve Act of 1891" originally contained no provision whatsoever for forest reserves.[37] The fateful Section 24 clause that authorized the President to set aside forest reserves—reserves which later evolved into our huge system of national forests—was added as a last minute rider by a House-Senate conference committee. It was never referred back to its originating committees—a technically illegal procedure.[38]

The forest reserve clause had more peculiarities than its irregular addition to a bill during a conference committee. Most importantly, the clause lacked any provision for appropriating funds or for management of any forest reserves the President might create. Without any legislative provision for the management or funding of the forest reserves, what was supposed to be done with them was a total mystery. In practice, that meant the reserves were absolutely closed, that no one was authorized to even set foot within them, much less make any kind of use of them.

As things turned out, the forest reserves seriously affected the western rancher. Within the tree lands were also grazing lands, including some of the choicest. The forest reserves created by Presidents Harrison and Cleveland by 1896 alone covered some 20 million acres, much of it valuable grazing land upon which ranchers had already established prior rights. As more reserves were created, especially the 21-million acre "Washington's Birthday Reserves" created by President Cleveland on that commemorative day in 1897, western ranchers demanded access to the grazing lands within them, but the question of whether stock would be permitted on the ranges was not even addressed until late in the 1890s.[39]

The effect was to tie up vast areas of the public domain so that no one could settle them, cut firewood or construction timber on them, hunt game on them, mine them for minerals, or even build transportation roads through them. The spreading pattern of forest reserves across the western map hemmed in new settlers and ranchers.

Conservationist calls for the creation of forest reserves had been common for twenty years. Conservationists of the era were well connected to northeastern financial moguls. The northern core financiers themselves were aware that there are two ways to monopolize any resource: One, get all of it for yourself that you can; two, keep anybody else from getting what you can't. Eastern capitalists had held a dominant position in the West for years, in cattle, mining, and railroads. They were adept at lobbying and at getting others to lobby for them—the Crédit Mobilier scandal of 1867-1872 demonstrates that the eastern capitalists knew how to turn federal land legislation to their advantage.

For example, the very first large reserve of any kind in America, Yellowstone National Park, was lobbied into existence in 1872 by an eastern capitalist. It was Northern Pacific railroad entrepreneur Jay Cooke who financed the Washburn Expedition of 1870 to explore Yellowstone. He also made sure that Nathaniel P. Langford—his agent—was on the expedition to recommend that the attractions not be divided up into small private claims, but made into a park for all to enjoy. The only practical access for visitors would be on the Northern Pacific railroad. The Washburn expeditioners lobbied Congress during the winter of 1871-72, again financed by Jay Cooke and the Northern Pacific. Jay Cooke and the Northern Pacific gained a federally protected monopoly on Yellowstone National Park and the access to it.[40]

HOW THE WEST WAS SUBJUGATED

The course of the first 20 million acres of forest reserves was a rocky one. After six years of bungled efforts to legislate proper uses for these reserves, Congress finally passed legislation authorizing management and appropriations for them. What is now called "The Organic Act," like the 1891 Forest Reserve Act, was sneaked in as an amendment to another bill, the Sundry Civil Appropriations Act, signed by President McKinley on June 4, 1897.[41] The amendment stated that the purpose of the reserves was for watershed protection and timber production. Lands primarily valuable for mineral ores and agriculture were to be excluded from forest reserves. In the reserves settlers were allowed to cut firewood, fencing, and building material, and mining and prospecting were specifically authorized within their boundaries. But grazing or access for stockmen to the range in reserves was not even mentioned.

The protests from western ranchers were heard in Washington. At the base of this protest was the prior appropriation doctrine. Less than a month after the new law passed, the General Land Office promulgated regulations that allowed grazing in the reserves. There was, however, no explicit recognition in the regulations of prior appropriated water rights or customary occupancy range rights. Nonetheless, a major hurdle in recognizing those rights in the forest reserves had been crossed.

The fact that easterners dominated most vital decisions in the western lands generated a long-smouldering sectional hatred that has never died, as former Colorado Governor Richard D. Lamm noted in *The Angry West*:

> By the end of the nineteenth century, the West was living precariously at the colonial end of a new American mercantile system. Every aspect of its industry was controlled and exploited by Easterners and eastern institutions, leaving little profit for those who lived on the land itself. Every aspect of its life was controlled by a federal government that viewed Westerners as wards of Washington and their land as a colonial empire. Sporadically, an increasingly angry West fought back—but the labor strikes and the Populist convulsion of the 1890s achieved little. As the century ended, only one thing was certain. Westerners had settled the land, dug it, planted it, and died on it.
>
> But they had not owned it.[42]

The beginning of the new century brought with it an unexpected turn in federal policy. The Public Lands Commission of 1903, an unpaid commission composed of men in the executive departments to investigate the public lands and to give recommendations for their management, studied the range and essentially sidestepped the private property idea hoped for by generations of ranchers. The commission suggested "something formally like private property:" government control of grazing through a permit system. It recommended the official creation of federal grazing districts.

These districts were to embody principles much like the ranchers' well-established arrangement that granted the owner of private "base" land an exclusive permit to use nearby federal range in large blocks. Preference to grazing permits was explicitly conditioned upon (1) customary occupancy of the federal land and (2) ownership of adjacent or nearby private land that was of necessity water-bearing. But within the reserves hope was now diminished of ever gaining full title to grazing lands for the private rancher. Prior appropriated water and land rights were swept under the rug by bureaucratic language asserting that ranchers had no preemptive right good against the United States, but only a prescriptive right good against rival contenders for use of the range, and that the landlord had given only permission, not a right, to graze its lands. The growing bureaucracy in Washington still made policy and ruled the land, but only by ignoring basic rights of the rancher.

The Forest Service, created in 1905, imposed regulations that provided for range permits and charges for use on the forest reserves, officially renamed "national forests" in 1907.[43] Western ranchers challenged the practice all the way to the U.S. Supreme Court and lost decisively in 1911 when two key cases, *Light* v. *United States* and *United States* v. *Grimaud* were decided in favor of the government's right to impose grazing permits and fees.[44] But neither case addressed the underlying possessory interests of the rancher in federal lands or the prior appropriation water rights upon which they were based. This is a loose string still dangling to this day, waiting for someone to pull it and unravel the now gigantic fabric of centralized government control.

THE TAYLOR GRAZING ACT

Congressman Edward T. Taylor of Colorado introduced a bill in 1934 to extend grazing regulations to all federal lands, not just the national forests. It proposed a Division of Grazing in the Department of the Interior, leaving some administrative features from the Forest Service approach, but giving far more decision-making power to local advisory boards in the grazing districts. The membership of advisory boards was elected by permittees. Preference was extended explicitly to established stockmen, land owners, settlers, and owners of water rights. Owners of base land obtained exclusive permits for specific allotments of federal land. Administration was supposed to be decentralized, not concentrated in Washington, as many ranchers had complained of the Forest Service. Congress passed the Taylor Grazing Act on June 28, 1934.[45]

The Taylor Grazing Act reflects the ranchers' continued determination well into this century to eventually obtain clear title to the federal rangelands they had long struggled to win by informal property arrangements that they hoped would be later recognized by the government. Ranchers supported the Taylor bill because it gave them more local control and appeared to offer the hope of eventually obtaining title to the land. Even Franklin D. Roosevelt's irascible empire-building Interior Secretary Harold L. Ickes bowed to that hope. He approved the clause in the first sentence of the Taylor Grazing Act that reads: ". . .in order to promote the highest use of the public lands *pending its final disposal*, the Secretary of the Interior is authorized. . ." (italics mine). Although the terms of that "final disposal" were left vague, it satisfied the ranchers at the time.

THE ERA OF REGULATION

The heritage of the Forest Reserve Act and the Taylor Grazing Act today is a vast federal land empire consisting of one-hundred-fifty million acres of land under the jurisdiction of the Bureau of Land Management (created July 16, 1946 by the merger of the General Land Office and the Taylor-Act's Grazing Service) and one-hundred-twenty million acres of land under the Forest Service.[46]

Anti-western federal policy did not go away. It continues today under the guise of environmental protection. As Richard F. Bensel stated in *Sectionalism and American Political Development: 1880-1980,* "most major environmental policies directly retard the development of natural resources in the West and, thus, indirectly stimulate the exploitation of similar resources (where they exist) in the East. Strip-mining restrictions, leasing for off-shore drilling on the Continental Shelf, and legislated prohibitions on the private use of western land in the public domain all fall into this category."[47]

As the political clout of commodity industries diminished in favor of the rising urban sector during the 1960s, a new movement appeared, partly as an expression of urban, mostly eastern power. Environmental lobbying groups such as the National Wildlife Federation, the Audubon Society, the Sierra Club, and the newer organizations such as the Friends of the Earth and the Environmental Defense Fund gained increasing influence with Congress and the budget process. Their combined membership of about five million and their combined annual budget during 1988 was more than $150 million and formed a substantial power and pressure bloc.[48] If the agencies wanted environmentalists' political support, the agencies would have to follow the environmentalists' agenda.

The West struggles under onerous federal enactments lobbied into existence by environmental groups funded largely by eastern capitalists and their corporate foundations.[49] The Wilderness Act of 1964 alone has eliminated more than 90 million acres of the West from commodity production at staggering lost opportunity costs.[50] The National Environmental Policy Act of 1969 has imposed Environmental Impact Statement (EIS) requirements upon hundreds of economic projects that were delayed and subsequently had to be abandoned.[51] The range livestock industry in 1974 was severely threatened when the Natural Resources Defense Council (NRDC), a public interest law firm with strong ties to northeastern money and politics, brought a lawsuit demanding that 212 site-specific EIS plans be drafted by the Bureau of Land Management for its grazing lands. *NRDC* v. *Morton* (388 F. Supp. 829) was designed to reduce livestock grazing on federal lands by 50 percent. Although NRDC won its case, the number of EIS plans was reduced to 144 and cattle reductions were minimized.

The federal livestock ranges have shown a decrease in production in the last few decades—not because of decreased biological productivity of the land itself, but because of regulatory overkill.[52]

One of the most serious symptoms of this federal dominance is the relationship of the individual rancher to the federal agencies when disputes arise about the rancher's grazing rights. If a rancher has a conflict with the Forest Service or Bureau of Land Management, he has restricted access to the court system. He must first "exhaust his administrative remedies." That means the rancher must often face a lengthy administrative appeals process controlled by officers of the very agency with which he has the dispute. The cost of defending himself sometimes becomes prohibitive. Loss of private property without compensation is often the result.

Access to justice for the rancher is not the only problem. The high cost of centralized federal regulation has stopped much mineral exploration and production. Many of the small free-lance prospectors and miners who have historically been responsible for many of our mineral discoveries are no longer operating. They can not afford the cost of complying with the massive regulations which apply to them.[53]

The amount of timber available for harvest from our federal lands has dropped dramatically, not because the timber is not there and capable of sustained-yield production, but because onerous environmental restrictions prevent its harvest.[54]

The dollar cost of this continuing restraint on the federal land economies of the western states is formidable. The pervasive inefficiencies of government management of the federal lands are only one restraining factor, but an important one. The Forest Service, for example, spent almost $2 billion in 1980, while its lands generated fewer than $1 billion in minerals, timber, grazing, and other revenues.[55] The cost of preparing Environmental Impact Statements and fighting environmentalist lawsuits, along with the cost of salaries, benefits, and pensions to the Forest Service bureaucracy makes economic use of the federal lands impossible.[56] This results in billions of dollars worth of American goods and services being withheld from our economy while we import products from foreign nations—and at the same time export our jobs.

In the 1980s, even when the Reagan administration promised privatization and deregulation, performance lagged far behind policy. Ironically, it was the ranchers' split estate interest in the federal lands that scuttled the 1982 Reagan proposal to privatize by sale some 35 million acres, or 5 percent of the public lands. The Reagan administration discovered that you can't sell what you don't own. As the President's Commission on Privatization lamely admitted:

> A year later, in July 1983, the entire effort was unceremoniously canceled. It had foundered on the existence of de facto property rights. Over 50 years or more of continued use, ranchers had acquired a de facto property right to graze on particular parcels of public lands—rights even bought and sold in the market on occasion. As a practical matter and whatever the legalities, such informal rights cannot be taken from their holders without the government providing fair compensation. In such instances, the best strategy for achieving privatization may be to give (or to sell cheaply) to the holders of political property rights some portion—or conceivably even all—of the rights to the government property being divested. This strategy had not been followed in the 1982 planned sales.[57]

The Reagan administration's failure to understand the split estate thus effectively foreclosed its options in conducting a rational privatization of appropriate portions of the federal lands.

CRISIS

Ranchers often own the water rights, springs, wells, stock watering ponds, water piping, irrigation ditches and rights of way, grazing permits, roads, fences, and corrals on federal lands. The commanding fact is this: Many ranchers already own the predominant interest in the split estate federal lands.

At the same time, environmental groups such as the National Wildlife Federation, the Sierra Club, and the Wilderness Society have raised the battle slogan *Livestock Free by '93!* They intend to eliminate all livestock grazing from federal lands by 1993.[58] They deny that any rights whatever belong to the rancher. They insist that grazing fees be raised to "market value," which they know is a mere ploy to bankrupt ranchers and remove them from the range. A typical environmentalist argument goes like this:

> A 1985 appraisal demonstrated that while the 30,000 ranchers who have grazing permits on 307 million public acres were paying $1.35 per "animal unit month" (the cost of grazing one cow for one month), the average rate for leasing comparable private land was over $6.[59]

It's a wonderfully shrewd play on public ignorance: They assert that ranchers are getting away with a subsidy from the taxpayer of almost five dollars per cow each month. They further assert that the rancher should be denied this subsidy by forcing him to pay "fair market value."

The flaw in the environmentalists' argument is this: There is no such thing as private land comparable to federal grazing permit land.

When private range is leased from a private owner for $6 per AUM (animal unit month), the lease includes the private owner's water rights, which could constitute a large fraction of the value of the leased range—$3 or $4 per AUM. The environmentalists argue the federal land rancher should pay $6 per AUM so his grazing fee can be "comparable" with private grazing lease rates. But most federal range permittees already own the water rights based on the prior appropriation doctrine. Why should they be made to pay $3 or $4 per AUM for something they already own?

Likewise, when private range is leased from a private owner for $6 per AUM, the lease includes the private owner's infrastructure—the roads, fences, corrals, etc., which makes up another $2 or $3 per AUM. But most federal range permittees already own that infrastructure. Why should they be made to pay $2 or $3 per AUM for something they already own?

Then too, when private range is leased from a private owner for $6 per AUM, the lessee may be leasing the private owner's trucks and cowhands to move the cattle around to fresh forage, which makes up another $1 or $2 per AUM. A federal range permittee must provide his own trucks and cowhands to move the cattle around. Why should he be asked to pay $1 or $2 per AUM for something he has already paid for?

Environmentalists choose to ignore that many ranchers already own most of the split estate in their federal lands.

The value assigned by some banks and Internal Revenue Service estate audits to the actual residual federally-owned value of a grazing permit AUM is approximately 60 cents. By this powerful argument, even a grazing fee of $1.35 per AUM is more than twice the real value of federal range. The current grazing fee is higher than the actual value of the range.

The fee would have to be higher than the actual value of the range in order to pay the salaries of the many federal employees (which private range owners do not hire), plus their retirement pensions, health coverage, and dental plans. It would have to be higher than the actual value of the range in order to pay the enormous costs of supporting wild horses and burros that consume forage and water at rancher expense. It would have to be higher than the actual value of the range in order to pay the enormous costs of fighting environmentalist lawsuits that challenge grazing plans wholesale at a costs of millions of dollars in lawyers' fees. And even with the subsidy from ranchers, the Forest Service spends $2 billion in taxpayers' money to generate $1 billion in revenues.

The taxpayer is not subsidizing the rancher. The taxpayer is subsidizing an inefficient bureaucracy. The rancher, like other taxpayers, is subsidizing the bureaucracy, the environmental movement, and the legal profession. Again: Many ranchers already own the predominant interest in the split estate federal lands.

The time has come to recognize those rights.

Chapter One Footnotes

1. Congressional Research Service, *The Major Federal Land Management Agencies: Management of our Nation's Lands and Resources*, Document 87-22 ENR, Library of Congress, Washington, January 14, 1987, p. 6.
2. The labyrinthine story of American wildlife law is well told in *The Evolution of National Wildlife Law*, prepared for the Council on Environmental Quality by the Environmental Law Institute, Government Printing Office, Washington, 1977.
3. Echo Bay Mines, *Annual Report 1987*, Denver, p. 17.
4. Nevada: *Nevada Statutes* 1925, c. 201, sec. 2.; California: *California Statutes* 1856, p. 54.; Colorado: *Colorado Constitution*, Article 6, 5. For an extensive discussion of these and other water laws see Wells A. Hutchins, *Water Rights Laws in the Nineteen Western States*, vol. 1, Department of Agriculture Miscellaneous Publication 1206, Government Printing Office, Washington, D.C., 1971.

5. The text of the Taylor Grazing Act may be found in 48 Stat. 1269. For pertinent examples of state laws see *Laws of Texas*, 1866, Sess. II, pp. 187-188; *Laws of Wyoming Territory*, 1873, Sess 3, p. 225; *Laws of Montana Territory*, 1872, Sess. 7, p. 287. The Organic Act of 1987 is in 30 Stat. 34.

6. For a thorough description and detailed history of the Concession Policy Act of 1965 see Don Hummel, *Stealing the National Parks: The Destruction of Concessions and Park Access*, Free Enterprise Press, Bellevue, 1987.

7. The overall topic of the politics of American natural resource decision making is expanded upon in a series of probing essays in John G. Francis and Richard Ganzel, eds., *Western Public Lands: The Management of Natural Resources in a Time of Declining Federalism*, Rowman & Allanheld, Totowa, New Jersey, 1984.

8. Robert A. Caro, *The Years of Lyndon Johnson: The Path to Power*, Random House, New York, 1981, pp. 12-13.

9. For a full discussion of the 98th meridian water barrier, see Walter Prescott Webb, *The Great Frontier*, Houghton Mifflin, Boston, 1952, and *The Great Plains*, Ginn & Co., Boston, 1931.

10. Cheyenne, Wyoming, *Leader*, May 8, 1868.

11. Alfred G. Cuzán, "Appropriators Versus Expropriators: The Political Economy of Water in the West," in *Water Rights: Scarce Resource Allocation, Bureaucracy, and the Environment*, Pacific Institute for Public Policy Research, San Francisco, 193, pp. 17-18. I have freely borrowed from Cuzán's argument for this section.

12. Charles W. McCurdy, "Stephen J. Field and Public Land Law Development in California, 1850-1866: A Case Study of Judicial Resource Allocation in Nineteenth-Century America," *Law and Society* (Winter) 1976, p. 236.

13. Hutchins here cites the case of *Jennison* v. *Kirk*, 98 U.S. 453, 457 (1879).

14. Hutchins, *Water Rights Laws in the Nineteen Western States*, p. 161.

15. William E. Colby, "The Freedom of the Miner and Its Influence on Water Law," published in *Legal Essays, in Tribute to Orrin Kipp McMurray*, pp. 67-84, 1935.

16. *U.S. Statutes At Large*, 39th Congress, Sess. I, Chap. 262, "An Act granting the Right of Way to Ditch and Canal Owners over the Public Lands, and for other Purposes," pp. 251-253.

17. Samuel C. Wiel, *Water Rights in the Western States*, Bancroft-Whitney Company, San Francisco, 1980, p. 12.

18. Key California cases include *Weaver* v. *Eureka Lake Co.*, 15 C. 271 (1860) which established the principle of beneficial use; *Yankee Jim's Union Water Co.* v. *Crary*, 25 C. 504, 85 Am.Dec. 145 (1864) giving the prior appropriation doctrine legal recognition; *Lux* v. *Haggin*, 8 P.C.L.J. 455 (1881), in which the court held that the federal water rights act of 1866 did not create any new right, but merely recognized and sanctioned preexisting rights, a vital reinforcement of the validity of informal property arrangements.

19. Key Nevada cases are *Barnes* v. *Sabron*, 10 Nev. 217 (1875)(reasonable beneficial use); *Jones* v. *Adams*, 19 Nev. 78, 6 Pac. 442 (1885)(custom and legislation); *Ormsby County* v. *Kearney*, 37 Nev. 314, 142 Pac. 803 (1914)(vested property rights).

20. *Jennison* v. *Kirk*, 98 U.S. 453 (1879).

21. *Walsh* v. *Wallace*, 26 Nev. 299, 67 Pac. 914 (1902).

22. Webb, *The Great Frontier*, pp. 254-259.
23. William D. Rowley, *U.S. Forest Service Grazing and Rangelands: A History*, Texas A&M University Press, College Station, 1985, p. 19.
24. *Itcaina* v. *Marble*, 56 Nev. 420 at 437.
25. Sworn statement by William Jones, rancher, Eddy County, New Mexico, April 10, 1917. Record Group 49, Unlawful Enclosure, Box 799, U.S. Archives. I am indebted to Professor Gary Libecap for this citation.
26. H. M. Taylor, "Importance of the Range Cattle Industry," *Annual Report* of the Bureau of Animal Husbandry, Washington, 1886, p. 316.
27. Ernest Staples Osgood, *The Day of the Cattleman*, The University of Chicago Press, Chicago, 1929, p. 185.
28. Ibid. pp. 186-7.
29. Elwood Mead, *Irrigation Institutions*, Macmillan Company, New York, 1903, pp. 28-29.
30. Richard Franklin Bensel, *Sectionalism and American Political Development: 1880-1980*, The University of Wisconsin Press, Madison, 1984, p. xix.
31. U.S. Congress, House, *Congressional Record*, March 27, 1888, p. 2463.
32. For a description of the Desert Land Acts of 1876 and 1877, see Benjamin Horace Hibbard, *A History of the Public Land Policies*, Macmillan Company, New York, 1924, reprinted by the University of Wisconsin Press, Madison, 1965, pp. 426-427. For a discussion of proposals to give rancher settlers 2,500 acre allotments, see *Disposal of Public Lands*, House of Representatives Report No. 778, 50th Congress, 1st Sess., Feb. 28, 1888, Washington, p. 6.
33. John W. Powell, *Report on the Lands of the Arid Region of the United States*, 45th Cong. 2d sess., 1878, H. Exec. Doc. 73, pp. 21-29.
34. The literature on the livestock industry is enormous. For histories of northern core depredations on earlier settlers see Ramon Frederick Adams, *The Rampaging Herd*, University of Oklahoma Press, Norman, 1959; Charles Bray, *Financing the Western Cattle Industry*, Fort Collins, 1928; Leslie Decker, *Railroads, Land and Politics*, Brown University Press, Providence, 1964; Matthew Josephson, *The Robber Barons: the Great American Capitalists, 1861-1901*, Harcourt, Brace, New York, 1934.
35. These northeastern core capitalists will be described in detail in Chapters Three and Five. I use the terms "northeastern core" and "northern core" interchangeably throughout the text, based on Bensel's analysis of sectional stress indicators.
36. *Buford* v. *Houtz*, 133 U.S. 618.
37. 26 Stat. 1095.
38. Harold K. Steen, *The U.S. Forest Service, A History*, University of Washington Press, Seattle, 1976, pp. 26-27.
39. Rowley, *U.S. Forest Service Grazing and Rangelands: A History*, pp. 29-30.
40. Hummel, *Stealing the National Parks*, pp. 6-9.
41. 30 Stat. 34-36.
42. Richard D. Lamm and G. Michael McCarthy, *The Angry West: A Vulnerable Land and Its Future*, Houghton Mifflin Company, Boston, 1982, pp. 12-13.
43. Forest Service. U. S. Department of Agriculture, *The Use of the National Forest Reserves*, Washington, D. C., July, 1905.
44. *Light* v. *United States* 220 U.S. 455 and *United States* v. *Grimaud* 220 U.S. 506.

45. 48 Stat. 1269.
46. *The United States Government Manual*, National Archives and Records Service, Washington, D.C., 1988.
47. Bensel, *Sectionalism and American Political Development:1880-1980*, p. 303.
48. The income of environmental groups is shown in their annual Internal Revenue Service tax exempt organization report Form 990. Copies of Form 990 returns do not include pages listing the source of the income, which is protected from public scrutiny by Act of Congress. Copies of each environmental group's Form 990 annual tax exempt income report must be obtained separately from the IRS regional office nearest the group's main headquarters, and are not available together at any single central point, such as the Reading Room of the IRS in Washington, D.C.
49. For tabulations of the enormous grants by eastern corporations and their foundations to environmental groups, see *Corporate Foundation Profiles*, The Foundation Center, New York, 1988, *The Foundation Grants Index*, The Foundation Center, New York, 1988, and *Environmental Law, Protection and Education*, The Foundation Center, Comsearch Grant Guides series #22, New York, 1989.
50. A detailed study of typical wilderness impact can be found in *The Adverse Economic Impact of Wilderness Land Withdrawals in Elko County, Nevada*, George F. Leaming, Ph.D., Western Economic Analysis Center. Presented at the Second Annual National Wilderness Conference, Reno, Nevada, April, 1989.
51. For a popular account by a distinguished scholar, see Bernard J. Frieden, *The Environmental Protection Hustle*, The MIT Press, Cambridge, 1979. For a scholarly account see Ellen Frankel Paul, *Property Rights and Eminent Domain*, Transaction Books, New Brunswick, 1987.
52. For a thorough discussion of these issues, see the chapter on Grazing Rights in Gary Libecap, *Locking Up the Range: Federal Land Controls and Grazing*, Pacific Institute for Public Policy Research, San Francisco, 1981, p. 45ff.
53. For a discussion of the evils of overregulation, see *Rights and Regulation: Ethical, Political, and Economic Issues*, edited by Tibor R. Machan and M. Bruce Johnson, Pacific Institute for Public Policy Research, San Francisco, 1983.
54. K. A. Soderberg and Jackie DuRette, *People of the Tongass: Alaska Forestry Under Attack*, Free Enterprise Press, Bellevue, 1989.
55. Thomas M. Lenard, "Wasting Our National Forests—How to Get Less Timber and Less Wilderness at the Same Time," *Regulation*, July/August 1981.
56. Sterling Brubaker, ed., *Rethinking the Federal Lands*, Resources for the Future, Washington, D.C., 1984.
57. *Privatization: Toward More Effective Government*, Report of the President's Commission on Privatization, David F. Linowes, Chairman, Washington, D.C., March 1988, pp. 242-243.
58. *USA Today*, Editorial Faceoff: Who owns the range? "Give ranchers their rights to rangeland," by Ron Arnold, May 3, 1989, p. 10A.
59. Ibid. "Don't give away the range; hike the rent," by David Alberswerth.

2

GENEALOGY OF THE FEDERAL LANDS

Today's storm over rangelands has been shaped by more than two centuries of contention over the federal lands. As we saw in Chapter One, the forces at work in that conflict span several dimensions. Sectionalism is one of the more unexpected.

Sectionalism is one of the principal agents that molded America's present federal land system. Growing numbers of historians—Richard Bensel among the foremost—are asserting that sectionalism has been and remains *the* dominant influence on the American political system. In this chapter we will pick up their theme—that sectional competition grounded in a geographical division of labor between the economically advanced northern core and the underdeveloped southern and western periphery drives American politics—and apply it to our discussion of the western rangelands.[1]

The most significant and enduring northeastern core/western periphery conflict has centered around the federal lands. The fact that such lands exist at all is the fundamental source of contention. As U.S. Interior Department economist Robert H. Nelson noted in analyzing the Sagebrush Rebellion and the privatization movement of the early 1980s, "Federal ownership of vast areas of western land is an anomaly in the American system of private enterprise and decentralized government authority."[2] The President's Commission on Privatization also remarked: "The public lands constitute about one-third of the land area of the United States, a huge federal domain of ownership that is hard to reconcile with the reputation of this country as a citadel of reliance on markets and the private sector."[3]

A sectionalist interpretation of Nelson's "historic anomaly" would begin with this fundamental assumption: At some time during the development of the West, the northeastern core decided to keep private property, and thus effective economic power, out of the hands of all potential competitors in the newly admitted states, and, with its concentrated political influence, was able to freely manipulate national policy makers to accomplish their purpose by making private property in the West as scarce as possible. This assumption suggests that today's federal lands *resulted* from sectional conflicts and are an instrument of northern core political power.

The northeastern core could attain its hypothetical objective by two complementary routes: One, by gaining control over all the western lands within its grasp, and two, by preventing western settlers from gaining control over the rest. If this hypothesis is supportable, three questions must be answered: Did northeastern core interests in fact gain control over all the western lands they possibly could? Did they prevent western settlers from gaining control over the rest? And if so, how did they obtain the manipulative power to do it?

THE AMERICAN SYSTEM

This conjecture deserves detailed scrutiny, so let's begin at the beginning. It hardly needs emphasizing that the United States was founded—if not developed—upon the belief that protected private property rights were an essential ingredient of a proper system of government. And, of course, private property rights are at the crux of our hypothesis about the non-private federal lands. As historian Richard Schlatter noted in *Private Property: The History of an Idea:*

> After the Declaration of Independence, most of the thirteen states adopted new constitutions, prefaced by Bills of Rights which included the natural right of property. The most famous, the Virginia Bill, declared that 'all men are by nature equally free and independent, and have certain inherent rights, of which, when they enter into a state of society, they cannot by any compact deprive or divest their posterity; namely, the enjoyment of life and liberty, with the means of acquiring and possessing property, and pursuing and obtaining happiness and safety.' Massachusetts followed with the declaration that 'all men are born free and equal and have certain natural, essential, and unalienable rights; among which may be reckoned the right of enjoying and defending their lives and liberties; that of acquiring, possessing, and protecting property; in fine, that of seeking and obtaining their safety and happiness.' From this time on, American states acquired the habit of proclaiming property a right of nature. Even in the twentieth century, when American political theorists had abandoned the theory of natural right, makers of constitutions continued to paraphrase the original Bills of Virginia and Massachusetts. A declaration of the natural right of property became one of the traditional elements in a state constitution.[4]

The British jurist Sir William Blackstone, upon whom the American Founding Fathers relied heavily, wrote that there is no principle more closely associated with a free society than private land ownership.[5] During revolutionary times John Adams saw private property as the most important single foundation of human liberty. He wrote:

> The moment that idea is admitted into society that property is not as sacred as the Laws of God, and that there is not a force of law and public justice to protect it, anarchy and tyranny commence. Property must be sacred or liberty cannot exist.[6]

Indeed, long before anyone could envision a United States on the American continent, it was the promise of private property, private land ownership—farms, businesses, homes—that drew early colonists to the shores of America more than any other single factor.

Even such a religious individualist as Roger Williams, who was banished by the Massachusetts court in 1636 for defying the authority of the state over his religious conscience and for asserting that the king's patent to the colonists conveyed no just title to the land, which should have been bought from its rightful owners, the Indians, was arguing for private property.

Private property in land was the lure over the centuries that brought countless millions to our shores. The underlying basis of America's historic East/West sectional conflict was the simple equation—no less true now than then—that *land equals power*. Land is the basic commodity out of which all other commodities gain their value; we live on it, we get our food from it, we base industries upon it, we obtain rent from it—and even maritime industries gain their value from land-based infrastructure such as ports, docks, lighthouses, and shipyards.

Land equals power—economic, political, social. So the origin of our sectional conflict over rangeland lies in questions of our government's claim to property in land.[7]

GENESIS OF THE FEDERAL LANDS

In order to grasp the modern storm over rangelands, we must have in mind a clear answer to the question: Where did America's public domain come from in the first place? The short answer is that the first public domain lands were acquired by the federal government after the American Revolution from "Western Lands" claims belonging to the original states. There was no federal land in the thirteen original states.

The long answer is this: Under their colonial charters, seven of the thirteen original states—Massachusetts, Connecticut, New York, Virginia, North Carolina, South Carolina, and Georgia—had claimed a vast expanse of "Western Lands," as they were called, located generally between the Appalachian Mountains and the Mississippi River, lands that would one day become the states of Ohio, Indiana, Illinois, Kentucky, Tennessee, Michigan, Wisconsin, and parts of Minnesota, Alabama, and Mississippi.[8]

Once the Revolutionary War had settled the matter of who ruled the colonies, the question of who ruled the Western Lands assumed immediate and urgent importance to the victorious revolutionaries. After Cornwallis surrendered at Yorktown in October 1781, peace negotiations with the British were imminent, and the Founding Fathers all knew that the fledgling United States needed those lands for national security reasons. The language of the colonial charters was vague and the validity of the land claims uncertain, but at the peace table the Americans demanded a western boundary at the Mississippi River and in the 1783 Treaty of Paris, they got it.[9]

The new nation had gained independence for itself and recognition of its sovereignty over the Western Lands. But sovereignty is one thing and ownership quite another. Sovereignty is jurisdiction and ownership is control of property rights.

Actual ownership of these extensive territories belonged only to the seven claiming states at the beginning of the Revolution. Despite bitter disputes between the seven states over which one owned exactly what parcel and where the boundaries lay, they stood together on one point: the emerging federal government did not own the Western Lands.[10]

The six non-claiming states were jealous of these Western Lands and suspicious of their owners, a disunity in wartime that the insecure United States could ill afford. The non-claiming states doubted whether their shaky new confederation could really be "a Union of equals" with seven of the thirteen states holding such unequal territory. The six non-claiming states found themselves forced to levy heavy taxes to cope with the mounting war debt, but realized that the land-owning states might easily pay off their debts from the proceeds of Western Lands sales. Fearing that bankruptcy might accompany the end of the war, the New Jersey legislature asked how half of the new states could "be left to sink under an enormous debt, whilst others are enabled, in a short period, to replace all their expenditures from the hard earnings of the whole confederacy?"[11]

The war debt lay at the heart of the federal lands matter for a good reason: The Continental Congress had decided in September, 1776, after the first disastrous summer militia campaigns against the British, to establish an army of regulars to serve the duration of the war. It was clear that substantial rewards had to be offered to achieve the changeover from temporary to permanent enlistment. In addition to a $20 money bounty, Congress promised *land*, ranging from 100 acres for a private to 500 acres for a colonel. The land was to be provided by the United States, and since the federal government owned none, that meant the land—or the money to pay for it—had to come out of the states in the same proportion as the other expenses of the war. Land was part of the war debt.[12]

The bounty measures did not sit well with the non-claiming states. As historian Rudolph Freund explained of the 1776 land promises:

These resolutions had been contested hotly in Congress before they were passed, but their execution caused still greater concern. The New England states believed that the land bonus was not sufficient to overcome the aversion to enlistment; civilians and soldiers complained, it was said, of the long engagement as a contract of servitude. These states wanted, therefore, to raise the money bounty, especially for privates, and passed acts to this effect in their assemblies. Maryland protested that it had no land of its own and thought that it would have to buy land for bounties from other states. Its council of safety proposed to raise the money bounty to the Maryland line [army contingent] by $10 instead of offering land. This caused consternation in the Continental Congress, where it was feared that this precedent might break the back of all North America. Maryland was officially assured that it would not have to make good the bounty grants in its individual capacity, because it was the intention of Congress to provide bounty lands at the expense of the United States. But Maryland would not recant; and in order to mollify it as well as the New England states, Congress finally permitted enlistment for three years under the money-bounty system and reserved the land bounty to those soldiers and officers who signed up for the full duration of the war or until they were honorably discharged.[13]

Although the perilous revolution demanded all possible solidarity, recalcitrant Maryland even refused to sign the Articles of Confederation unless all Western Lands claims were abandoned by the seven claiming states. Expressing the feelings of the six non-claiming states, Maryland insisted that the Western Lands "wrested from the common enemy by blood and treasure of the Thirteen States, should be considered as common property."[14]

Common property meant *federal* property.

The claiming states, resisting the Continental Congress, refused to give up their Western Lands interests. Why should non-claiming states have equal benefit of such potential economic prosperity and political power? The Virginia Assembly offered a compromise in December, 1779, "to furnish lands out of their territory on the north west side of the Ohio river, without purchase money, to the troops on continental establishment of such of the confederated states as had not unappropriated lands for that purpose. . . ."[15] Maryland, a small non-claiming state, resented this offer from Virginia because it would make final consent to completely surrender the Western Lands under reasonable terms that much more difficult.

All disputants over the land issue clearly understood that the fate of the embattled young Confederation hung upon harmonious relations between all the thirteen states. Even so, the claimants held out. Practically in desperation, the Continental Congress, on October 10, 1780, pledged that all land ceded by the original states to the federal government "shall be disposed of for the common benefit of the United States, and be settled and formed into distinct republican States, which shall become members of the Federal Union, and have the same rights of sovereignty, freedom, and independence, as the other States."[16]

Despite this pledge and constant cajolery and rhetoric, it was not until a year and a half later, on March 1, 1781, that the first claimant gave in. On that date New York, in an act "to facilitate the completion of the Articles of Confederation," ceded its Western Lands to the federal government. On the same day Maryland signed the Articles after receiving promises that all seven claimants would eventually relinquish their Western Lands.[17] Finally the Confederation was ratified.

It didn't last. Sectional fights over the public lands were a good part of the reason. As soon as the preliminary peace accords were signed with the British in November of 1782, the soldiers of the Continental Line wanted the bounty money and bounty lands Congress had promised them. When the money and lands were not immediately forthcoming, mutinous enlisted men forced Congress to flee from Philadelphia to Princeton.

The officers of the Continental Line did not take part in the bounty lands rebellion, but rather sent a long list of demands to Congress in December of 1782, insisting upon immediate payment of arrearages and payment of a lump sum of money upon discharge. Congress under the Articles of Confederation had no power to tax, and could only recommend speedy action to the states. The states did nothing. The disgruntled officers did not wait.

In March, 1783, an anonymous officer at a New York conclave incited to open rebellion in the famous Newburgh addresses: ". . . while the swords you wear are necessary for the defence of America, what have you to expect from peace, when your voice shall sink, and your strength dissipate by division?. . . Can you then consent to be the only sufferers by this revolution, and retiring from the field, grow old in poverty, wretchedness, and contempt?" Then came the dire threat: ". . . the slightest mark of indignity from Congress now, must operate like the grave, and part you forever: that in any political event, the army has its alternative. If peace, that nothing shall separate you from your arms but death: if war, that . . . you will retire to some unsettled country, smile in your turn, and 'mock when their fear cometh on.'"

This last was an exhortation to refuse to lay down their arms until Congress granted their demands or, failing that, to abandon Congress as it had abandoned them, migrate to the West and begin a new semimilitary life together. Fortunately, General Washington himself unexpectedly appeared at a meeting of their delegates and talked them into an orderly appeal.

By April 1783, several officers involved in the Newburgh incident, including Brigadier General Rufus Putnam, had drawn up an elaborate proposal to Congress to create a new state for veterans on the Ohio. It was the beginning of the Ohio Company, one of many private land companies that subsequently attempted to act as intermediaries between government and people in the migration West.

Putnam's "army proposal" asked Congress for a long list of benefits, centering around the federal purchase of a new state for them in the Western Lands corresponding roughly to the present state of Ohio.

Putnam went on to emphasize that at least some of the officers held rather decided views as to the manner in which these new lands should be distributed and administered. They were, he declared, much opposed to monopoly and wished to guard against large patents granted to individuals; they hoped, therefore, that no grants would be made except by townships 6 miles square (or multiples), to be subdivided by the proprietor associates themselves and administered after the pattern of a New England town. This township principle was also to apply to private purchasers of land in the region.[18]

In this "army proposal" to create the New England-owned Ohio Company is found a prototype expression of northeastern intent to dominate the public domain lands. Its features are remarkable. It contained the anti-monopoly rhetoric that would be heard for a hundred years masking a covert intent to be the monopolist. It recommended a grid system of small holdings that would one day collide with the invisible water barrier west of thirty inches of rain and deny adequate land holdings to ranchers in the arid West.

According to the "army proposal," the Ohio Company, a corporation of New England veterans, would build a new community in the empty expanses of the West—but doing things their way: compact units with a distinctive rectangular grid system which settlers could enter only after the land was surveyed. The southern system, on the other hand, allowed pioneers total freedom to scatter where and when they liked and draw boundaries any shape they pleased. Had the southern system been adopted, there would likely be no federal land in America today.

But there was a rationale behind the "army proposal:" Eight years of war had rendered many officers of the Continental Army unfit to re-enter civilian life and they knew it. They had banded together in 1783 in the Society of the Cincinnati to preserve their *esprit de corps*—fully 90 percent of the signers of the Newburgh petition that led to the "army proposal" belonged to this order. George Washington was its first president, and supported their Ohio Company concept. The township system suited their needs exactly: It created little republics of convenient size, placing the civil and political power in the hands of those who owned the country, and through shares in the Ohio Company, the veterans would own the country. Prominent members of the Society of the Cincinnati pressed for the settlement of the West through the Ohio Company, which finally succeeded in executing a contract with Congress on October 27, 1787, obtaining a million and a half acres for the sum of one million dollars.[19]

Meanwhile, the most important Western Land relinquishment had come in the form of Virginia's Act of Cession of March 1, 1784. It is the most important because it became the model for all those that followed and it contained the crucial contradiction that was to generate endless turmoil over federal lands that lasts to this day. It stipulated that the ceded lands

> shall be considered as a common fund for the use and benefit of such of the United States as have become, or shall become, members of the Confederation or federal alliance of the said States, Virginia inclusive, according to their usual respective proportions in the general charge and expenditure, and shall be faithfully and bona fide disposed of for that purpose, and for no other use or purpose whatsoever.[20]

Neither Virginia nor Congress realized that with this language they were establishing two contrary principles for the public domain: first, to manage it as a "common fund" for the use and benefit of all, and second, to entirely "dispose of" it into new states. Managing the public domain as a "common fund" meant *retaining it in federal ownership*, at least until it could be sold to private settlers.

Virginia's Act of Cession, then, is the source of the idea that the federal government could retain the public domain, despite its effusive protestations otherwise. Thomas Jefferson in particular insisted that the federal government should sell all its vast domain to private owners, and then it should "never after, in any case, revert to the United States." Apparently no one at the time could foresee the mischief those two words "common fund" would one day work against the basic American system of private property, private enterprise, and decentralized government.

At the time of the Virginia Cession, it remained to be seen whether new states might enter the Union on an "equal footing" with the original states, and with their guaranteed rights of "sovereignty, freedom, and independence" intact while the federal government retained the public domain—and therefore control of the terms under which land could be opened for settlement.

Control of the public domain was the first effective power the federal government had in America: The Confederation Congress could not levy taxes on the states nor could it force them to deliver required national requisitions. But in the new public domain the federal government had absolute power, both jurisdiction and property rights in the soil. The Confederation had in its grasp what Jefferson wanted most to avoid: government ownership of the means of production.

In the public domain Congress could govern directly, it could raise needed revenue by sale of land, it could create new states. The public domain was the new government's greatest asset. It was the sole avenue available to centralized federal control. It was the first instrument of expanding federal power.[21]

The land-claiming states soon realized that fact, and many dragged their heels, endlessly putting off final cession of their Western Lands. Massachusetts ceded the last 34.5 million acres of her disputed lands in 1785, Connecticut ceded 25.6 million acres in 1786, South Carolina ceded 3.1 million in 1787, and North Carolina ceded in 1790 what is now the state of Tennessee—which never became a "public land state" because its entire area was already covered with privately owned land claims.

It is significant that the last cession of Western Lands to the federal government did not take place until April 24, 1802, long after the original Articles of Confederation had been abandoned and the Constitution ratified. On that date Georgia released its Western Lands claims—but only on four stringent conditions: First, that the United States was to pay Georgia $1.2 million from the sale of public lands in the cession (the only payment any state received for its Western Lands); second, the reservation of 500,000 acres to satisfy private claims against land in the ceded territory; third, extinguishment of Indian titles to certain portions of the cession; and fourth, agreement that all persons who actually settled within the territory ceded, "shall be confirmed in all grants legally and fully executed" prior to the date of cession.[22]

America's public domain arose from a feud started by six states over seven other states' claims to Western Lands that were finally ceded to the federal government by sectionalist political pressure under a fatefully flawed management assumption.

Today's federal lands *resulted* from sectional conflicts. But to see how they became an instrument of northern core political power we must follow the development of the new states that were carved out of the Western Lands—the public domain states.

CONFLICTING GOALS

Formal logic tells us that when the two premises of a syllogism contradict each other, any conclusion whatsoever follows, even nonsense. The evolution of the American federal land system may be regarded in that light. Although the federal government's intention expressed in the Virginia Cession of disposing of all the public domain was undoubtedly sincere, the new federal government did not resist exploiting the "common fund" of the public domain as a path to power.

Almost as soon as the first public lands came into existence (those ceded by New York in March of 1781) the Continental Congress proposed on September 5, 1782, to sell them to "discharge the national debt." This decision was true to the promise that public lands would be disposed of to settlers. But selling federal land rather than giving it away was a novel concept, something unknown under British rule.

Thomas Jefferson, whose dream of a nation of yeoman farmers required ample free land, at first opposed the idea of selling the public domain. But in 1784 even he incorporated the idea of selling federal land to meet public expenditures in the ordinance of that year. The government needed money. In the 1784 law the federal government made the first provisions to keep the public domain for itself.

As historian Daniel Feller noted,

> In early legislation for the West, Congress reaffirmed the dual doctrines of the Virginia cession, while clarifying its intention to retain ownership of the public domain even after the creation of new states. A congressional ordinance of 1784, prescribing a territorial government and a statehood process for the western regions, stipulated that new states "in no case, shall interfere with the primary disposal of the soil by the United States in Congress assembled," and also "that no tax shall be imposed on lands the property of the United States." The Northwest Ordinance of 1787, superceding the earlier legislation, repeated both prohibitions, but also provided that upon reaching a population of 60,000 each new state would be admitted "on an equal footing with the original States, in all respects whatever." By fiat, the ordinance proclaimed its own provisions as "articles of compact" between the existing and future states, to "forever remain unalterable, unless by common consent."[23]

In other words, the Ordinances of 1784 and 1787 meant that new states upon admittance to the Union would not receive title to the land within their borders: The federal government would keep all of the land in the state and sell it directly to private parties itself.

The federal land system was officially established by the Ordinance of 1785, the essential features of which have never changed. It follows closely the New

England pattern advocated by the Society of the Cincinnati and the Ohio Company. The public domain was to be properly surveyed prior to settlement in new territories, with townships six miles square, each township subdivided into thirty-six plots of a mile square (640 acres) numbered from one to thirty-six. Lands were to be auctioned off at a minimum price of one dollar an acre. Half of all land in a new territory could be sold in 640 acre sections, but the other half had to be sold in complete townships. One section, number 16, was reserved in each township "for the maintenance of public schools."[24]

This school reservation provision of the 1785 ordinance is worth comment: Congress had no explicit authority to subsidize education in either the Articles of Confederation or later in the Constitution. But there was little objection to the practice and the congressional school "donations" quickly became a permanent fixture in American land policy. Congress could demand that one-thirty-sixth of every new state be used as Congress saw fit. It would not be the last time that Congress gave itself additional power by embedding some popular provision into a land law. These original 1785 school land grants, as historian Feller caustically noted, "furnished a plausible precedent for similar grants for other purposes."

The landmark Northwest Ordinance of 1787 established a form of government for territories in the northern sections of the Western Lands and specified procedures for admitting them as states to the Union, which gave direction to the land disposal process.[25] However, the Northwest Ordinance did not include the southern sections of the Western Lands, a fact which generated some interesting consequences, which we will examine in due course.

The revenue from the first sales of public land was disappointing, another factor that encouraged the United States to retain the public domain. The initial surveys in eastern Ohio took two years to complete, and the auction at New York in 1787 produced only $117,108 (and that in depreciated securities) for 72,974 acres. The United States did not offer public lands for sale again until 1796, nine years later.

Part of the reason was the new Constitution of 1787. No federal land operations took place while it awaited ratification. Article IV of the Constitution granted Congress plenary power over the public domain, "Power to dispose of and make all needful Rules and Regulations respecting the Territory or other Property belonging to the United States; and nothing in this Constitution shall be so construed as to prejudice any Claims of the United States, or of any particular State." While this confirmed the authority already being exercised under the Articles of Confederation, the Constitution did nothing to clarify it. The new government inherited the state deeds of cession to the Western Lands, and Congress in 1790 and again in 1795 pledged to use land revenues "solely" for payment of the national debt.[26]

Another provision in the Constitution played a strong role in the disposition of federal land. The doctrine of states' rights is based on the Tenth Amendment: "The powers not delegated to the United States by the Constitution, nor prohibited by it to the States, are reserved to the States respectively, or to the people." Americans have always quarreled over what this meant. Alexander Hamilton and the Federalist Party favored a broad interpretation, which meant a strong central government deriving its authority from implied as well as enumerated powers contained in the Constitution. Thomas Jefferson, James Madison, and their followers, "strict constructionists," insisted that all powers not specifically enumerated in the Constitution were reserved to the states. The states' rights doctrine first appeared in the Kentucky and Virginia Resolutions of 1798, written by Jefferson and Madison. A further set of resolutions passed the next year by Kentucky and Virginia provided for state nullification of objectionable federal laws, but no other state adopted them. Despite this failure, the states' rights theory of government has been invoked by opponents of centralized federal power ever since.

While the states' rights doctrine was emerging, Congress in 1796 fixed the minimum land price at two dollars per acre, twice the price ordained in 1785, but offered sales on a credit basis, 5 percent down, 20 percent in forty days, and the balance in three equal annual installments, at 6 percent per annum interest. The basic sale size was reduced from an entire section (640 acres) to 320 acres in 1800 and down to 160 acres in 1804.

When Ohio entered the Union in 1802, Congress expanded the federal power once again through a land law. Ohio's emerging legislature did not relish the idea of having an entire class of citizens possessing property and at the same time paying no taxes—which is what happened because of the credit system: Since title to the land did not vest in the settler until he paid the final installment four years after his initial purchase, the state could not tax his land during that four years. Without the power to sell credit land for delinquent taxes—because it still belonged to the United States—Ohio felt its power to enforce tax collections had been rendered moot.

Congress offered the new state of Ohio an alternative: in return for their state constitutional convention refraining "by an ordinance irrevocable" from taxing any lands sold by the United States for five years from the date of sale, Congress would give Ohio some sections containing salt springs, donate one section in each township for schools (already reserved by the Land Ordinance of 1785), and spend 5 percent of all future Ohio net land proceeds to build roads to and through the state.

This gave Congress authority to subsidize transportation, another power not enumerated in the Constitution. Ohio readily agreed, requesting only that the road fund be divided, 3 per cent to be spent within the state, 2 percent on roads leading to its borders. Under these terms, Ohio entered the Union as the first public land state. The executive who signed the Ohio Enabling Act of 1802 into law was none other than President Thomas Jefferson.

In the law that created the state of Ohio we find the administration of Thomas Jefferson, a strict constructionist and champion of states' rights, committing itself and all future administrations to contracting internal improvements within the states. John Quincy Adams later asserted that the Ohio Enabling Act of 1802 coupled with the Louisiana Purchase doomed Jefferson's dream of a yeoman America.[27]

Six states entered the Union under terms similar to Ohio: Louisiana (1812), Indiana (1816), Mississippi (1817), Illinois (1818), Alabama (1819), and Missouri (1821). Kentucky (1792) never became part of the pubic domain because Virginia ceded its Kentucky land claim directly to the new state instead of to the federal government. Tennessee (1796) too never became part of the public domain because the territory was rapidly covered with private claims. By the time North Carolina ceded Tennessee to the federal government in 1790 there was virtually no unclaimed land left.

The Southern states of the Ohio model entered the Union under a new disability: Louisiana, Mississippi, Alabama, and Missouri, as noted above, were unfortunately not covered by the "articles of compact" in the Northwest Ordinance of 1787, and their enabling acts contained this harsh requirement of the respective state constitutional conventions:

> . . . that the said convention shall provide by an ordinance, irrevocable without the consent of the United States, that the people inhabiting the said territory do agree and declare, that they for ever disclaim all right or title to the waste or unappropriated lands, lying within the said territory; and that the same shall be and remain at the sole and entire disposition of the United States.[28]

The new states additionally had to renounce all taxing power over the public domain forever.

The sectional fracture lines in American politics began to propagate. Power was to remain in the northern core. The periphery was to accept inferior status and federal domination. By 1821 the federal government was well on its way to colonializing the West.

THE SECTIONALISTS EMERGE

After the War of 1812 had opened the floodgates of westward migration, millions of acres of western land lay open for settlement. The General Land Office, a bureau of the Treasury Department created in 1812 to administer the sale of the public domain, was swamped by immigrants. Its surveyors and auctioneers could not keep up with the pioneers who lunged headlong beyond the official frontiers despite an 1807 law forbidding trespass on public lands.[29]

Those who settled beyond the paperwork were technically deemed "squatters." Under the law, one could not (and still cannot) alienate land from the sovereign by adverse possession. Unless the sovereign, i.e., the federal government, agrees to recognize the claims of squatters, no amount of actual occupancy, land improvement, economic contribution or civic good can validate those claims. If the sovereign accedes, the process is called preemption. The squatter can then become the lawful owner of title to the land.

The concept of preemption rights is an important element in the story of today's storm over rangelands, so we will follow the trail of early American squatters with interest.   Squatters in the land rush after 1812 demanded preemption rights, which precipitated a political crisis. President James Monroe made a tactical mistake in trying to fight illegal settlement of the West by issuing a December 1815 proclamation warning squatters "who have unlawfully taken possession of or made any settlement on the public lands . . . forthwith to remove therefrom" or the army would forcibly eject them and the courts prosecute them for trespass.[30]

Monroe's action was not popular among westerners. As it turned out, the army could not enforce the proclamation and the pioneers just came back to their claims after the soldiers kicked them off. Congress soon authorized squatters to remain as temporary tenants if they obtained permission from local land officers. Very few obtained permission, but the General Land Office eventually caught up with the squatters and properly registered their claims, delivering fee title into their hands, which settled the matter peaceably. This preemption process became one of the primary methods by which settlers actually obtained most of the federal lands in the American West. The primary result of the Monroe-squatter clash was to reinforce western suspicions of eastern intentions to suppress western growth and power.[31]

By 1819 westerners numbered more than a third of the United States. Sectional antagonism filled the air with grumbling. Historian Feller depicted it most colorfully:

A Bostonian worried over the "influence, indefinable, but possibly terrible" soon to be exercised by "that *western world*." A new Kentucky congressman discovered that his Carolina and Georgia messmates "seem to think that the W[estern] C[ountry] will receive the requisite population to acquire the ascendency" in Washington. "I can assure them," he added, "that if the ability is ever possessed the power will be exercised."

The sight of farms and cities springing up in the wilderness inspired many Americans to heady visions of continental empire. . . . But in the Atlantic states, other effects of the western lure soon began to appear: abandoned farms, declining population, falling property values. "You, Sir, can't conceive of the anxiety and confusion that pervades all ranks of people in this section of country to remove to the Alabama," remarked one North Carolinian. "The consequence is that land here is deminishing in point of value, and the country is loseing many of its most enterprising and respectable inhabitants." A Virginia legislative committee lamented the state's sorry condition: "How many sad spectacles do her lowlands present, of wasted and deserted fields! of dwellings abandoned by their proprietors! of Churches in ruins!" A similar committee in North Carolina found it "mortifying to witness the fact, that thousands of our wealthy and respectable citizens are annually moving to the West . . . and that thousands of our poorer citizens follow them, being literally driven away by the prospect of poverty."[32]

The sectional issue of the day centered around the price of western land and the General Land Office practice of disposing of the public domain on credit under authority of a 1796 statute. Western immigrants wanted lower land prices and the eastern establishment wanted higher land prices. Western immigrants wanted to keep the credit system and the eastern establishment wanted to eliminate the credit system. Eastern manufacturers deplored the labor drain as the lure of easy land ownership drew workers westward, but westerners deplored the trade deficit because they had little to sell to the East. Eastern politicians complained about the loss of voters (and taxpayers) but westerners complained that the East was draining them of cash for land, needed goods and manufactures. The West developed a suspicion that the East, despite friendly rhetoric in public, secretly wanted to suppress its growth.

Again President Monroe aggravated the circumstances: In December 1817 he recommended that Congress increase the minimum price of public lands. The Illinois *Western Intelligencer* responded:

It cannot be possible, that this measure has been dictated in a spirit of jealousy of the rising importance of the western country. The president we believe to be too magnanimous for that. But certain it is, that such a measure if effected, will tend much to retard the rapid population which is now experienced daily. That the western country will ultimately acquire an important stand in the government, is evident, and no restrictions, nor obstacles which can be thrown in its way, will prevent, that effect from being sooner or later produced. The government, therefore, should aid in promoting this event, as it will tend to harmonise the two sections of the country. But if the western country finds a spirit of opposition prevailing in the national councils to its prosperity, when it shall acquire that stand which it will ultimately attain, a spirit of reciprocal opposition will perhaps be manifested, and give rise to those discontents which critics of our government have so often predicted. The government should pause before this step is taken.[33]

This thinly veiled threat of rebellion reveals the depth of western feeling about eastern suppression. Nor was it remarkable for its tone. Other publishers openly accused the East of economic and political conspiracy, particularly during the Panic of 1819 when the credit system collapsed and left the federal government holding twenty-two million dollars in unpaid land debts. Legislators began talking about ending the credit system. The *Albany Register* commented on the possibility that western settlers might have to forfeit the lands they could not pay for, warning that forfeiture might produce "a civil war, if not an ultimate dissolution of the federal compact."[34] The *Kentucky Reporter* in 1819 editorialized:

Nothing has been done to promote our local interests; and every scheme to advance our local prosperity and give to us a fair participation in the benefits of the union, has been thwarted or defeated. . . . The truth is, that the western states and territories and the population in them, are considered at Washington as *inferior* to the states and the people on the other side of the mountains. We are regarded as a sort of colonies or distant provinces, that may serve to swell the pride of empire, but which can by no means be considered as entitled to the same privileges with the Good Old Thirteen United States.[35]

Angry sectionalism that threatened to erupt into civil war was in full flower. Rufus King, a Senator from New York, expressed the covert eastern attitude when he wrote in a 1819 letter to a friend, "the demands and strength of the West are increasing daily and the vigorous decision and union of the old States decrease in a fully equal degree."[36] King wanted to eliminate the credit system on public lands and he wanted Congress to further regulate westward expansion. If the credit system continued, King warned, "the result will be to draw off our active population, to become purchasers of western lands, which are nominally sold, but which will in effect be acquired, without payment by the purchasers."

President Monroe, who himself was the object of western suspicion, once commented that Senator King revealed his hostility to the West on every land bill that came before the Senate.[37] On the credit issue King won: On March 2, 1821, the Relief Act ended the credit system but allowed settlers extended payments on their debt. It would take ten more relief laws and eleven years to eliminate the debt. And sectional hatred grew more virulent than ever.

SOME AMERICAN ARCHETYPES

The issues of sectionalism have from the very first involved a code language that put on a public face of solidarity between East and West, yet served to mask the true underlying forces of hostility and competition. Its beginnings were seen in the "army proposal" to Congress arguing against a monopoly in order to create a monopoly. The favored code language rhetoric after 1800 tended to cast economic questions in moral terms, fastening on two sharply contrasting American archetypes defined by historian Daniel Feller: the virtuous "actual settler," and the rapacious land speculator—who was seen, of course, as a monopolist.

As Feller has noted, the actual settler personified the Jeffersonian ideal. "Honest, hardworking, loving liberty, he wanted only a small farm from which to support his wife and children. His poverty matched his virtue. Tragically unable to provide for himself on the wornout, high-priced lands of the East, he drifted westward in search of a home. Though he might fall on hard times and petition Congress for relief, his sufferings were always traceable to impersonal forces, or to the evil machinations of others. His character was pure, his patriotism invincible."[38]

The speculator, on the other hand, was a parasite who contributed nothing to the nation, who grew rich off the labors of others and monopolized large tracts of land by holding them for sale at obscene profits.

In fact, when the moral terms are set aside, most actual settlers were indistinguishable from speculators. Farmers by the thousands bought land for resale on credit along with the eastern agents.

The rhetoric of these archetypes struck a deeply responsive chord in the American public of the day and during most of the nineteenth century. Public opinion held the two archetypes so concretely that they personified the opposites of good and evil as solidly as Christ and Satan. Such a firmly held public belief made a perfect screen for hidden intents: Any eastern politician who wanted to restrict western growth and economic prosperity had only to invoke the "actual settler" and berate the "rapacious speculator" and he could justify almost any piece of legislation regardless of its actual content.

It became commonplace for advocates from the Atlantic states to voice a concern for the welfare of western "actual settlers" that they did not feel. This habit stayed with Congress well into the twentieth century. Westerners were not fooled, but without hard evidence to back up their suspicions they could not make a convincing exposé. Western correspondence and literature from the early 1800s onward is full of comments on eastern "reformers" who were suspected of entertaining covert motives to inhibit the growth of the West.[39]

Using another piece of archetypal code language, eastern politicians worked very hard during the 1820s to paint the western states as "pampered children," opulently endowed with public lands at the expense of their older brother states in the East. It was part of an effort to persuade Congress to give the eastern states a portion of the school donation lands awarded to each new western state. The issue came to a head in 1821 when Maryland state senator Virgil Maxcy drew up resolutions asserting the right of the Atlantic states to a share of federal school land donations in western states. Maxcy's proposal said that the old eastern non-public land states had a *right* to the lands of the new western public land states, and even put down on paper the exact acreages due the "excluded states" to show eastern state legislatures how much revenue could be obtained by the eastern states if Congress passed the proposal. Eastern states rushed to support Maxcy's bill.

The proposal was brought before Congress, where Senate Public Lands Committee chairman Jesse Thomas of Illinois (a western state at the time) opposed the idea of entrusting western lands to eastern states. Thomas insisted that the new states had in fact paid for their school donations by agreeing not to tax federal lands until five years after they were sold. Thus, Thomas said, the school grants, "though distinguished in common parlance by the name of *donations*, were in fact sales bottomed upon valuable considerations, in which the new States surrendered their *right* of sovereignty over the remaining public lands, and gave up the whole amount which might have been received in taxes before such lands were sold, and for five years thereafter."[40] Thomas was pointedly noting that if it were not for the admission compacts dating from the Northwest Ordinance of 1787 and based on the contradictions of the Virginia cession, the federal government would not own the public domain at all.

Thomas' statement punctured the eastern code rhetoric. The East might perceive the West as a favored sibling, but the West did not see itself that way. The federal land system was just another burden dumped on the pioneer's already strained back. The school grants certainly didn't make up for the loss of sovereignty over the public domain, to say nothing of the piles of cash drained away to Washington through the General Land Office. In fact, the compacts of admission had simply taken advantage of the powerlessness of the new states to allow the old states to extort concessions that were nothing short of blackmail.

The Maryland proposition failed in Congress, but it exacerbated the sectional hatreds already engendered by the crisis of 1820 over the ending of the credit system. Again, early in American history, East and West came to the brink of civil conflict over public land issues.

LAND AND POLITICAL PARTIES

As America matured into the Jacksonian era, the nation that had been marked for decades by weakly organized political action and virtually insignificant political parties grew into the exact opposite. By 1840 the Whigs and Democrats had built highly structured organizations with clear-cut alternative ideologies. Land policy was a central feature in both parties.

The Democrats held onto the fading Jeffersonian ideals of frugality, simplicity and virtue. They were strict constructionists of the Constitution who upheld limited, decentralized government. Their land policy was based upon Jefferson's commanding vision of a nation of yeoman farmers that countenanced business and industry, but did not promote them. They offered cheap land and easy preemption, fostering individual freedom, personal initiative and the least government consistent with health, safety, and public morality. Their rhetoric was strictly agrarian. The Democrats' republic was strictly agricultural.

The Whigs, on the other hand, saw government as an active instrument of progress and improvement. They were broad constructionists who wanted commerce and industry as much as they wanted agriculture. Their land policy emphasized distribution of land revenues to the states for education, internal improvements and Negro colonization, but they saw the public lands as a capital fund needed to endow their program of social betterment. They wanted strict controls on migration, not halting it, but preventing it from draining the population of the East. Their rhetoric was anti-western, casting the frontier in terms of lawlessness and disorder, damning the squatter "not as a heroic pathfinder but as a renegade and outcast, a depredator of resources, and a trespasser upon the Indians."[41]

During these years of conflicting party pressures on land policy, nothing much really changed. A crucial proposal that could have changed history entirely—to cede public lands to the states in which they were situated—was recommended during the congressional session of 1831-1832 by the Secretary of the Treasury, but voted down by the Public Lands Committee in both houses.[42] The minimum size of a purchase unit dwindled from 160 to 40 acres. Credit sales for $2.00 an acre were abolished in favor of cash sales at a minimum price of $1.25 per acre. One important change with substantial and lasting effects was the legalization of preemption. In 1830 and again in 1834, preemption received the cachet of Congress as a legally guaranteed right, no longer a crime: settlers on lands not yet offered for sale could later purchase them at the prevailing minimum price.

CONCESSIONS TO THE WEST

In 1841 Congress passed a law that donated 500,000 acres of public lands directly to nine western states and promised the same gift to all new states that might be admitted into the Union.[43] The need for improved transportation between the East and the rapidly developing Middle West provided the incentive: The lands were to be sold and proceeds used for the promotion of "internal improvements." States might select lands within their boundaries "in such manner as the legislature thereof may direct and in parcels of not less than 320 acres." Proceeds were to be "faithfully applied to objects of internal improvement, namely: roads, railways, bridges, canals, and improvement of water courses, and draining of swamps." All routes were to be "free for the transportation of the United States mail and munitions of war, and for the passage of their troops without the payment of any toll whatever."[44]

The states of Ohio, Indiana, Illinois, Alabama, Missouri, Mississippi, Louisiana, Arkansas, and Michigan were to be given these grants for internal improvements, but previous grants such as Ohio's five percent road grants were subtracted first. Thus Ohio and Indiana received no additional land, Illinois got 209,000 acres and Alabama 97,000 acres. The other named states received their full 500,000 acre allotment, as did those admitted between 1841 and 1889 (except for West Virginia and Texas): Florida, Iowa, Wisconsin, California, Minnesota, Oregon, Kansas, Nevada, Nebraska, and Colorado.

The 1841 Act also gave ten percent of the net proceeds from public land sales to the several states over and above what each state was entitled to by its articles of compact upon admission—and it granted preemption rights to all settlers, as had the Acts of 1830 and 1834.

But the internal improvement law of 1841 did not mark an end to anti-western sectionalism, or even a lull in the battle. During debate on the bill, Congressman Henry Alexander Wise of Virginia blasted the measure, bitterly complaining that it was "especially inequitable between the new states and old." Representative Robert Charles Winthrop of Massachusetts argued against the new states that "the contemplated cession would be a fatal dowry to them, as well as a measure full of injustice to us." Predictably, Democrats opposed the bill and Whigs supported it. It won in the House by eight votes—116 to 108.[45]

The clash of Whigs and Democrats during the Jacksonian era seems to have ended in a balance of forces if not stalemate. But it was not to last. The year 1850 marked the watershed. It was in that year that journalist John Louis O'Sullivan gave us the coinage, "Our manifest destiny is to overspread the continent allotted by Providence for the free development of our yearly multiplying millions." Also in 1850 Indiana boasted the nation's seventh largest population; Illinois was eleventh.

Railroads, which had begun to spring up in 1830, had by 1850 ripened into an independent transportation system, no longer merely the adjunct to canals and river traffic. At mid-century the railway system of New England was essentially complete, and the Middle and South Atlantic states had developed sufficiently so that directions of future growth could clearly be seen. The trunk lines all terminated in New York, which guided trade to the northeastern core. The West traded with the East more than with the South, which established in America a closer union West/East than West/South.

The year 1850 stands as the precise watershed in congressional land policy: The Illinois Central Act of 1850 began a new epoch of railroad grants. A respected historian once said of this law:

> The terms of the grant provided that alternate sections of land for six miles on each side of a line of railroad extending from Chicago, Illinois, to Mobile, Alabama, and from the mouth of the Ohio River to Dubuque, Iowa, should be given to the states within which public land lay, i.e., Illinois, Alabama, and Mississippi, and through which the road should pass. This land was to be handled by the state legislatures and sold, the money to be used in building a railroad. The law did not say whether the states should build and own the road or not, but it was understood that a private road might receive the proceeds as a subsidy. On this latter plan the states very promptly acted.[46]

Here in 1850, in the first railroad land grant Act, is the maturation of the principle of federal aid to states for transportation which had its roots in the Ordinance of 1784 and the Ohio Enabling Act of 1802. Jefferson the strict constructionist and champion of states' rights had taken part in forming these early laws. Ironically, their logical conclusion, the historic series of railroad land grant laws of the second half of the nineteenth century, once and for all sealed the doom of his bucolic vision of an agrarian America. The railroads were not owned by yeoman farmers and they did not promote an agrarian economy: They were owned by eastern corporate capitalists intent on bringing the Industrial Revolution into being.

The federal lands continued to serve as an instrument of expanding northern core political power.

Interior Department economist Robert H. Nelson may have underestimated the actual historic role of public land conflict when he wrote, "Federal ownership of vast areas of western land is an anomaly in the American system of private enterprise and decentralized government authority." Today's federal ownership of vast areas of western lands is not so anomalous after all. The very first public land states, Ohio, Indiana and the others, suffered much the same problem; the vital difference is that most if not all the public domain land in the states admitted between 1800 and mid-century was eventually sold into private hands while today's federal estate is a permanent capital asset of the United States government not available to the public for purchase.

By 1850, the federal land question had become a central focus of sectional conflict: Shall new western states be carved from the public domain slave or free? What rights of sovereignty shall new western states have over the land within their boundaries? Shall the East continue to colonialize the West? All the seeds had been planted that would one day erupt into that most sectional of American sectionalist conflicts: the Civil War.

Chapter Two Footnotes

1. The sectionalist interpretation of American history was given fresh impetus and respectability by Richard Franklin Bensel in *Sectionalism and American Political Development: 1880-1980*, The University of Wisconsin Press, Madison, 1984. Many historians still regard the sectionalist interpretation as a fading anachronism harking back to Frederick Jackson Turner's disciples, but Bensel argues convincingly with modern evidence that sectionalism "has been and remains the dominant influence on the American political system," with which I concur. Another modern sectionalist analysis can be found in Daniel Feller, *The Public Lands in Jacksonian Politics*, The University of Wisconsin Press, Madison, 1984.

2. Nelson, Robert H., "Why the Sagebrush Revolution Burned Out," *Regulation*, May/June 1984, p. 27.

3. *Privatization: Toward More Effective Government, Report of the President's Commission on Privatization*, David F. Linowes, chairman, Washington, D.C., March 1988, p. 242.

4. Richard Schlatter, *Private Property: The History of an Idea*, Rutgers University Press, New Brunswick, New Jersey, 1951, pp. 188-9.

5. Sir William Blackstone, *Commentaries on the Laws of England.* 4 vol. The *Commentaries* were first published between 1765 and 1769. I have used the twelfth edition, edited by Edward Christian, Professor of the Laws of England, Cambridge, London, 1794.

6. John Adams, *Works*, edited by C. Francis Adams, vol. IX (1854), Little, Brown, Boston, 1850-56, p. 560.

7. Many historic commentators have emphasized the importance of land and property in land. See Rousseau's *Discourse on Political Economy* in its entirety and Montesquieu's *The Spirit of Laws*, particularly Books V, VIII, XVII, and XXVI on property in general; and the classic economic works, particularly the long chapter on "The Rent of Land" in Adam Smith, *An Inquiry into the Nature and Causes of the Wealth of Nations*, first edition, London 1776; and Chapter XXXIII, "The Modern Theory of Colonisation," in Karl Marx, *Capital: A Critique of Political Economy. Volume 1, The Process of Capitalist Production*, translated by Samuel Moore and Edward Aveling, edited by Frederick Engels, International Publishers, New York. Modern dissent on land ownership as it relates to an environmentalist "land ethic" is discussed in J. Baird Callicott, "American Indian Land Wisdom? Sorting Out the Issues," *Journal of Forest History*, Vol. 33, No. 1, January 1989.

8. All this is detailed at length in Benjamin Horace Hibbard, *A History of the Public Land Policies*, Macmillan Company, New York, 1924; reprinted by the University of Wisconsin Press, Madison, 1965, p. 8 ff.

9. Treaty of Paris of 1783, *Treaties and Conventions Concluded Between the United States of America and Other Powers Since July 4th, 1776*, Government Printing Office, Washington, 1889.

10. Payson Jackson Treat, "Origin of the National Land System under the Confederation," in *The Public Lands: Studies in the History of the Public Domain*, edited by Vernon Carstensen, University of Wisconsin Press, Madison, 1962, pp. 7-14.

11. *Journals of the Continental Congress, 1774-1789*, Worthington Chauncey Ford et al., editors, 34 vols., Washington, 1904-1937, Government Printing Office, Wash-

ington, vol. 11, p. 650.

12. The land bounty was offered in the Resolution of Sept. 16, 1776, in Ford, *Journals of the Continental Congress, 1774-1789*, vol. 5, pp. 653-655. For a vivid picture of the importance of this measure, see the letter of George Washington, Sept. 2, 1776, to the President of Congress in Peter Force, ed., *American Archives*, series 5, 3 vols., Washington, 1848-1853, vol. 2, pp. 120-121.

13. Rudolf Freund, "Military Bounty Lands and The Origins of the Public Domain," in *The Public Lands: Studies in the History of the Public Domain*, pp. 17-18.

14. *Maryland's instructions to her delegates in Congress*, Dec. 15, 1778. See Johns Hopkins University Studies, III, No. 1, p. 26.

15. William Waller Hening, *The Statutes at Large; Being a Collection of All the Laws of Virginia, 1619-1792*, 13 vols., Richmond, 1809-1823, vol. 10, p. 559.

16. This story is told in the crucial documents collected by Thomas Donaldson, *The Public Lands*, Government Printing Office, Washington, 1884, p. 64.

17. Resolution of the Congress of the Confederation, Oct. 10, 1780, *Journals of Congress*, VI, pp. 146-147.

18. Freund, "Military Bounty Lands," p. 24.

19. The details of the Ohio Company's formation are in *The Records of the Original Proceedings of the Ohio Company* edited by Archer Butler Hulbert in *Marietta College Historical Collections*, I and II.

20. Donaldson, *Public Domain*, pp. 68-69.

21. I have freely adapted this section from Daniel Feller, *The Public Lands in Jacksonian Politics*, University of Wisconsin Press, Madison, 1984, pp. 4-6. Feller is another scholar who like Bensel asserts that sectionalism was and is the dominant force driving American politics.

22. See Hibbard, *A History of the Public Land Policies*, pp. 12-14.

23. Feller, *The Public Lands in Jacksonian Politics*, p. 6.

24. For a minutely detailed treatment of these events, see Paul Wallace Gates and Robert W. Swenson, *History of Public Land Law Development*, Public Land Law Review Commission, Washington, 1968; or Payson Jackson Treat, *The National Land System, 1785-1820*, E. B. Treat & Co., New York, 1910.

25. Treat, "Origin of the National Land System under the Confederation."

26. For both acts see *The Public Statutes at Large of the United States of America*, Little and Brown, Boston, 1845-1848, vol. 1, p. 144 and p. 435.

27. Adams to Alexander Everett, May 24, 1830, and September 18, 1831, *American Historical Review* 11, January 1906, pp. 340-343.

28. See the enabling act of each of these states: *The Public Statutes at Large of the United States of America*, Vol. 2, pp. 641-643; vol. 3, pp. 289-291, 348-349, 489-492, 545-548.

29. *U. S. Statutes At Large*, Ninth Congress, Sess. II, Ch. 46, March 3, 1807, "An Act to prevent settlements being made on lands ceded to the United States, until authorized by law," pp. 445-446.

30. James D. Richardson, *A Compilation of the Messages and Papers of the Presidents, 1789-1897*, GPO, Washington, 1896-1899, vol. 1, p. 572.

31. See Malcolm J. Rohrbough, *The Land Office Business*, Oxford University Press, New York, 1968.

32. Feller, *The Public Lands in Jacksonian Politics*, p. 15.

33. Nathaniel Pope in the Kaskaskia (Illinois) *Western Intelligencer*, January 21, 1818.

34. Albany (New York) *Register*, reprinted in the *National Intelligencer*, August 11, 1819.

35. Kentucky *Reporter*, as reprinted in the St. Louis *Enquirer*, December 4, 1819.

36. King to Christopher Gore, February 11, 1819, and April 9, 1820, in *The Life and Correspondence of Rufus King*, ed. Charles R. King, New York, Putnam's Sons, 1894-1900, vol. 6, pp. 207-208, 212.

37. Feller, *The Public Lands in Jacksonian Politics*, p. 33.

38. Ibid., p. 29.

39. James Campbell to Allen Trimble, February 11, 1819, in *Old Northwest Genealogical Quarterly*, vol. 10, July, 1907, p. 261. Rufus King to Christopher Gore, July 17, 1819, in *King Correspondence*, vol. 6, p. 228.

40. *American State Papers: Documents, Legislative and Executive, Public Lands*, vol. 3, pp. 410-411, 496.

41. Feller, *The Public Lands in Jacksonian Politics*, p. 190.

42. Hibbard, *A History of the Public Land Policies*, p. 229.

43. *U. S. Statutes At Large*, Twenty-seventh Congress, Sess. I. Ch. 16, Sept 4, 1841, "An Act to appropriate the proceeds of the sales of the public lands, and to grant preemption rights," pp. 453-456.

44. *Laws of the United States*, vol. X, pp. 157-158.

45. *Congressional Globe*, 27th Congress, 1st Session, p. 156.

46. Hibbard, *A History of the Public Land Policies*, p. 245.

3

SECTIONALISM AND THE FAR WEST

To understand the 1862 Homestead Act that came to dominate western land ownership patterns we must examine the period surrounding the Civil War. Keep in mind, the American Civil War resulted in chaos unparalleled in our nation's history. We lament, and justifiably so, the loss of nearly sixty thousand lives in the ten years of Vietnam conflict. Our Civil War, however, resulted in the slaughter of six hundred thousand men, ten times the number killed in Vietnam. This horrible toll mounted in just four years, from 1861 to 1865.[1]

Prior to the outbreak of the Civil War, there was intense competition between the North and the South to persuade the newly forming states and territories of the West to embrace their particular political persuasion. President James Monroe signed the Missouri Compromise in 1820, providing for the admission of Missouri into the Union as a slave state but prohibiting slavery in the rest of the northern Louisiana Purchase territory. Even though the act was supposed to quiet the controversy on the nation's frontier, it had just the opposite effect. Emotions flamed, generating barbarity on both sides exceeded only by the conflagration it foreshadowed.[2] I mention these things to underscore the political climate in which the Homestead Act philosophy developed.

The earliest wave of Forty-niners who followed the siren call of Sutter's gold to California came primarily from Georgia, Alabama, and other southern states.[3] They were soon followed, of course, by gold miners from around the world. But southerners were among the first owners of prior appropriation rights on the federal lands reaching from the Sacramento Valley up to the Sierra Nevada.

As the Civil War approached, the North realized it must gain firm control of the mineral wealth of California's Mother Lode as well as the Comstock Lode discovered in western Utah Territory—now Nevada—in 1859. The Comstock, a huge silver-rich fissure in the eastern flank of the Sierra Nevada near Lake Tahoe, was touted as the largest deposit of precious metals ever discovered. The South could not be allowed to seize that treasure by the agency of southern claimants, but mining law had to remain intact if continued production was to be assured. The presence in the region of so many southern immigrants and Copperheads—northerners with southern sympathies—made northern interests uneasy.

The Mormons were another source of apprehension. Northern leaders well remembered the efforts of the Church of Jesus Christ of Latter Day Saints under Brigham Young to establish a State of Deseret in 1849. "Deseret," according to the Book of Mormon, means "honeybee," a symbol of industrious devotion. As historian Hubert Howe Bancroft stated, "An immense tract of country was claimed, extending from latitude 33° to the border of Oregon, and the Rocky Mountains to the Sierra Nevada, together with a section of the territory now included in southern California, and the strip of coast lying between Lower

49

California and 118°30′ of west longitude. The seat of government was to be at Salt Lake City. . ."[4] That strip of Deseret coast, incidentally, ran from the present Mexican border to Los Angeles. A Constitution was adopted after a drafting session on March 4, 1849, elections were held for a legislature, an executive and a judiciary on March 12, and all officials took the oath of office on July 3—the State of Deseret was established.

The United States had other plans for the California section claimed by Deseret. But there was a problem. The Californios—American settlers—in Lower California, as the southern section was still called, had resisted martial law after the United States claimed the area in 1846. Secessionist elements continually called for a division with the establishment of a slave state in Lower California and a free state in Upper California.

President Zachary Taylor requested that Brigham Young and other Mormon leaders consider a "temporary amalgamation of the states of California and Deseret, in order to avoid possible difficulties on the slavery question. It was agreed that a memorial should be drawn up, asking for a convention of all the people of Upper California, both east and west of the Sierra Nevada, for the purpose of consolidating the two states in one that should include all the territory acquired from Mexico."[5] The Mormons agreed, but the governor of California rejected the idea out of hand.

It is worth noting that at this time the Californians, angered by the inaction of Congress on their statehood application, followed the example of the Mormons by framing a constitution of their own and demanding admittance to the Union. The next year was a busy one. The federal government granted statehood to California, rejected the Mormon petition to enter the Union as the State of Deseret, and hastily organized the Territory of Utah, including almost all of what is now Nevada except for the southern tip, and, of course, eliminating all Mormon claims to California and seaport access.

Mormon leader Brigham Young was appointed governor of the Utah Territory. Mormons had by then settled much of the region, to a limited extent even in what is now Nevada. Friction between the Mormons and the federal government began almost at once. Petitions for statehood were repeatedly denied by Congress. Federal judges, often incompetent, were assigned to Utah Territory and generated clashes over interpretation of the law. Promised federal appropriations were not forthcoming. United States land law was applied to Utah Territory and the Mormons found themselves merely squatters in their own land. The friction grew into open conflict.

In 1857 President James Buchanan declared Utah Territory in a "state of substantial rebellion," removed Governor Brigham Young, and directed U. S. Army troops to pacify Utah. Young prepared for war, recalling to Salt Lake City most of the few Mormons who had settled in the western portion of the Utah Territory. Mormon guerillas harassed westward-bound troop supply trains, but the "Utah War" was settled peaceably before it grew into widespread bloodshed. Nevertheless, ill feelings thus engendered on both sides lingered to the eve of the Civil War. The South realized that if it could neutralize or win over the Mormons, it could strengthen their cause. The North realized it too.

In 1859 the Comstock Lode brought miners from California into the western portion of the Utah Territory. The Comstock Lode served as a catalyst that brought the new western land policy together. It was so rich, both North and South realized, that control of the deposit could well be decisive in the outcome of the war. The North moved immediately for control against a local population with strong southern sympathies.[6] This had to be done delicately.

The first move was to ease the Comstock Lode away from control of the Mormon community based in Salt Lake City. Carson County was created to surround and contain the mineral wealth. Then came the first calls for a separate Territory of Nevada. Once the Carson County buffer zone had been created between the lode and the remainder of Utah Territory it was necessary to protect this vast wealth from any attempted armed takeover. An overt movement of federal troops to the Comstock was viewed as an invitation to conflict with southern interests. The United States government was well aware that many Nevada immigrants were southern veterans of the Mexican War of 1846-1848 well experienced in military matters.[7]

A war in 1860 with the Paiute Indians provided the excuse for the establishment of Fort Churchill on the Carson River near the Comstock. There is some evidence that the Paiute War was provoked by northern interests to create local support for establishment of the fort and stationing of federal forces in the area. Whatever the cause of the Paiute War, after the Indians defeated the local militia with heavy loss of life in the first battle of Pyramid Lake, resistance to stationing federal troops in the area decreased.[8]

A strong anti-Mormon feeling was fostered by pro-Union interests, accompanied by stronger demands for the separation of Western Utah from the Utah Territory in a proposed new Territory of Nevada. A "Declaration of Cause for Separation" appeared in mid-1859, stating "That a long train of abuses and usurpations on the part of the Mormons of eastern Utah, towards the people of western Utah, evinces a desire on their part to reduce us under an absolute spiritual despotism. Such has been our patient sufferings, and such is now the necessity for dissolving all political relations that may have connected us together, and we deem it not only our right, but also our duty, to disown such a government, and such a people, and to form new guards for our future safety."[9] Thus more political tension exacerbated the growing war paranoia and the rank lawlessness of raucous wide open boom towns such as Virginia City.

The sectional breach was hopelessly riven by the election of Republican Party candidate Abraham Lincoln as president in the fall of 1860. When war erupted at Fort Sumpter on April 12 the next year, Congress hastened to create Nevada as a separate territory, securing it to the Union from the South and from the Mormons. But, despite all pledges of loyalty from Nevadans, Congress was not really sure it had secured the new territory from the resident Copperheads and southern immigrants. A huge secret society of Southerners and Copperheads known as the Knights of the Golden Circle was abroad in the land from coast to coast—estimates placed its membership at over 300,000—inciting to rebellion and arming its furtive allies. Its "golden circle" was to be a vast loop of slave states extending from the South to California, thence southward including all of Mexico, Central America, part of South America and back up to encompass all the Carribean Islands including Cuba.[10] Surely its agents were at work in Nevada. Nevada Territory tried to reassure Washington by passing a law depriving those who were disloyal from voting at elections.[11]

At this crucial juncture, a Kentucky lawyer named David S. Terry, a justice of the Supreme Court of California who was appointed a federal judge by President of the Confederacy Jefferson Davis, led a band of southerners from Calaveras County into the Comstock and fortified it so that Nevada would be part of the Confederacy. The judge and his volunteers could not hold their defenses and nothing came of the Terry rebellion.[12]

Then all Nevada brought itself under suspicion of disloyalty: The Constitutional Convention of 1863, designed to bring Nevada into statehood, broke down over numerous personal quarrels about nominations to office, quarrels that split the Union party. As a popular history of Nevada notes, "The split was a serious one, and before the Convention had made its nominations a formidable movement in the Union ranks had arrayed itself in hostility to the Constitution, to which the secession element in the Territory immediately joined hands."[13] The people overwhelmingly rejected the proposed state Constitution at the polls January 19, 1864. That was tantamount to declaring for the Confederacy.

Nevertheless, the recalcitrants of the vastly rich Nevada Territory were given a second chance by Congress, another election was held, the people approved the state Constitution, and Nevada was rushed into statehood in mid-1864. Congress gave Nevada statehood even though it lacked sufficient population to qualify.[14] Congress deemed gold and silver more important than the technicality of population requirements in this crisis. In addition, Nevada's two new loyal Unionist Senators—James W. Nye and William W. Stewart, both from New York—were needed if Lincoln was to win the vote on the Thirteenth Amendment. The new governor, H. G. Blasdel, was also a staunch Union man.

Washington never forgot that Nevada was rife with southerners—about 30 percent of the total population. Along with the southerners, Copperheads in unknown numbers owned extensive prior appropriation rights to gold and silver claims, to water, and to livestock grazing land. Rebels were disenfranchised in the new constitution. A Union man at the state constitutional convention had grumbled that Nevada Territory "had been made a place of refuge for the scum and off-scourings of the rebel states."[15]

There was worse for Washington to worry about. Separatists in California agitated for a completely autonomous western nation, not part of the Union, not part of the Confederacy, but allied with the Confederacy.[16] Separatists including the Knights of the Golden Circle promoted a republic comprising the present states of California, Oregon, Washington, and parts of the Mexican states of Sonora and Baja California, to be revolutionized by filibustering expeditions. William Walker, the strange man from Nashville, Tennessee, who conquered Nicaragua in 1855 only to be forcibly deported by United States Marines in 1857, was a filibuster of the Knights of the Golden Circle.[17] There were many more like him in California all through the Civil War. They had money, they had arms and ammunition, they had a grand plan for California—and for an immense empire.

Nevada found itself saddled with Union war objectives. The 160 acre maximum ownerships provided for in the Homestead Act assured that only a very small portion of the state would pass into private ownership, thereby assuring continued federal control of its mineral-rich mountain ranges.

This was an intensification of federal land policy that began with the admission of Ohio in 1803, and that westerners had viewed for half a century as an "extra burden piled upon the already great travails of pioneering," a policy by which "cartloads of cash [were] drained from the West through the land offices." The federal government had "taken advantage of their political powerlessness to extort concessions that came close to blackmail."[18] Nevada then, became a state only in a cosmetic sense, but for all practical purpose remained a territory ruled by the federal government. It set the pattern that would affect all states entering the Union after 1864.

HOMESTEAD ACT SHAPED BY CRISIS

Congress viewed the withholding of the land base from private ownership in the western states as an expedient necessary until the Civil War crisis had ended. In 1861-'62 most victories in the new war went to the South. Southern troops stood on the doorstep of Washington. Congressmen attended the sessions with weapons at hand, not knowing when they might have to fight for their lives on the Capitol steps. It was in this highly charged atmosphere that debates over western settlement policy took place and gave rise to the Homestead Act of 1862.[19]

Let's take a closer look at the Homestead Act to better see what it was intended to do.[20]

First, the Homestead Act required potential settlers to be citizens of the United States not in rebellion against the United States. This effectively ruled out the pro-southern settler. Second, the Homestead Act allowed for a one hundred sixty acre maximum claim. This greatly limited the danger of one individual obtaining enough of the resource base to become politically or economically dangerous to the Union cause.

The obvious problem with the Homestead Act as an effective tool for eventual divestiture of the retained federal lands to private ownership was the nature of the West itself. The Congressman from Ohio or Vermont viewed one hundred sixty acres as a huge farm. To him the Homestead Act seemed very generous. In reality, a very small portion of the West could be economically farmed in tracts as small as one hundred sixty acres. Climate, elevation, and topography greatly restricted the kinds of crops that could be grown. A major portion of these arid lands were adaptable only to livestock grazing. Viable livestock grazing required not acres but sections of land. The 1862 Homestead Act did not provide for conveyance of title on tracts larger than one hundred sixty acres.

From the outset a new pattern of land use developed in the West. Home-steaders settled on their hundred sixty acre parcels and grazed their cattle on the adjacent unappropriated federal lands, the public domain. The private home-stead served as a base, the neighboring unclaimed public domain served as their range. To secure their claim on a certain piece of range they claimed the water arising on that portion of range.[21]

The early ranchers obtained legal recognition of their water rights on the federal lands as a result of controversy over miners' water rights. In 1865 northeastern interests introduced proposals in Congress to expropriate the mines on the public domain from the miners, and operate and sell them to raise money for the war debt. It was a Reconstruction-era excess that failed. Western members of Congress, led by Nevada Senator William Stewart, ground the campaign to a halt. Stewart was not a man to be trifled with: as a young lawyer he had been known to resort to the six-shooter if the law failed.[22] The Senate calmed down when Stewart spoke reason to them, and western indignation culminated in the Act of July 26, 1866, expressly confirming the rights of miners and appropriators that had theretofore been recognized only tacitly. It recog-nized the customs and usages that had grown up on the public lands under state and territorial sanction. Specifically concerning water, it provided

> That whenever, by priority of possession, rights to the use of water for mining, agricultural, manufacturing, or other purposes, have vested and accrued, and the same are recognized and acknowledged by the local customs, laws, and the decisions of courts, the possessors and owners of such vested rights shall be maintained and protected in the same.[23]

This act had the effect of separating the non-navigable waters from the federal lands, giving them to the states. This began the classic split estate land ownership pattern that came to dominate much of the West.

The states were then able to convey the use of water to private parties who met state requirements. The race to claim waters on the federal lands was intense as settlers scrambled to establish property rights.[24] It was widely assumed by citizens and the government alike that post-Civil War politics would act to convey range land titles to those who now owned the water and had used the range beneficially. It was difficult for anyone to foresee the obstructionism that would soon prevent the clearing of title.[25]

THE GENERAL LAND OFFICE

One of the first major roadblocks was the General Land Office, the agency responsible for divestment of the federal domain. The General Land Office, transferred from the Treasury to the Interior Department in 1849, administered the Homestead Act. Most people during the last half of the 19th century could go through life without ever meeting a federal government employee. The few federal employees there were held a somewhat privileged position. Their incomes were good and their power considerable. The Land Office employee had a strong incentive to preserve his job by whatever means that came to hand.

The General Land Office employee significantly increased his income each time he processed a homestead filing. Officials at each land office received fee and commission payments for validating and processing claims, supplementing their incomes up to a three-thousand dollar limit. Agents also had power of discretion over who qualified for a homestead.

Any talk of conveying fee title to the vast acreages of range that had now become vital parts of ranching operations was a threat to the General Land Office employee's job and security. If he could continue to drag out the land settlements in one hundred sixty acre parcels, his fee income, agency budgets, and long term career were secure.[26]

These General Land Office personnel, being a part of the Department of Interior, were in a position to influence Congress on western land issues to a degree far out of proportion to their numbers.

THE ROBBER BARONS

Other interests displayed the same obstructionism as the General Land Office for very different reasons. The northeastern capitalists also wanted to prevent or slow the passage of the federal domain into the hands of westerners in order to control the West themselves.[27]

In 1855, nineteen New Yorkers were millionaires. William B. Astor, son of John Jacob Astor, Stephen Whitney, James Lenox, assorted Rhinelanders, Schermerhorns, and Goelets, William Aspinwall, Alexander T. Stewart, Peter Cooper, George Law, and Cornelius Vanderbilt.[28] Although Vanderbilt held the lowest place in the 1855 millionaire register, by the time he died in 1877 he was the richest man in America. Vanderbilt was the first great mogul of U.S. capital and industry, accumulating a fortune over $100,000,000. Even though he entered the railroad field late in life, gaining control of the New York Central only ten years before he died, the Commodore forged his rail holdings into a vast empire. Cornelius Vanderbilt pioneered the business style that proved once and for all that there is a difference between capitalism and free enterprise:

> He appreciated the values of monopoly. Although he thought the public should pay aplenty for the benefits of monopoly, he believed those benefits should be real. Both Rockefeller and Carnegie, and other successful industrialists, possibly profited by his pioneering example. He knew how to buy, or at least sway, the acts of judges, of public officials of all sorts, and of legislatures. He knew how to get extra-curricular help from the federal government, including regiments of marines when wanted. Old Vanderbilt even set the pattern of elegance and display followed, though greatly expanded, by so many of his class who came later.[29]

Neither Vanderbilt nor any other American built a really great fortune until the Civil War. Chicago's meat-packing moguls did not rise to riches until military demand fattened their market. New England's great boot and shoe industry marched forward with the rate of war mobilization. The gun business bolstered a wobbly iron and steel industry. The gun running business stimulated ship building. But most importantly, the war acted as a hothouse that forced the growth of the railroads.

It was the advent of the railroad after the Civil War that opened up the Mountain West to heavy settlement and at the same time opened up the eastern market for hardy livestockmen from San Antonio to Cheyenne. The advent of the railroad also brought together many of the grasping robber barons who were to become the northeastern core cattle capitalists of the West. The same sectionalist hatred that ignited the Civil War lingered decades afterward, allowing predatory northeastern core robber barons to go unchecked in their domination of the West.

The idea of a transcontinental railroad had appeared as early as 1832, but growing North-South hostility prevented the project from getting under way. Neither side would approve a route that might give an advantage to the other.

Not until 1862, when the withdrawal of southern representatives and senators gave Congress a more unified voice—except for the "Peace Democrats," i.e., Copperheads—did a bill pass authorizing construction of the Pacific railroads. A route was approved based upon political and military considerations. The Union Pacific was authorized to build a line westward from Omaha, Nebraska, to the California-Nevada line, while the Central Pacific was to build simultaneously eastward from Sacramento, California. The Union Pacific found a way to skirt Mormon Utah by passing north of all major settlements. The Promontory Point route was strictly political: a practicable passage through Salt Lake City had been discovered in 1850. The Central Pacific's terminus at Sacramento was firmly in the hands of loyal Union supporters. Their route proceeded protectively from the Mother Lode of the Forty-niners directly to the Comstock Lode of Nevada, and thence eastward.

The law provided for a grant of land in behalf of the Union Pacific and Central Pacific roads. Each road received odd-numbered alternate sections of land for twenty miles on each side of the road—a swath forty miles wide—and the railway companies were required to allow preemption by settlers on any land remaining three years after completion of the road, which should be sold to them for not more than $1.25 per acre. The width of the grant provided both protection and control for the railroad, and therefore for the federal government. If politics should demand control over who settled where with what access to transportation, or if rebellion should demand rapid troop movement, both could be readily commanded.

Along with the land came money, a loan of from $16,000 to $48,000 for each mile of track laid. The land grant was made directly to the railroads instead of to the states to subsidize the companies, giving the railroad capitalists direct claims on federal money.[30]

When the Union Pacific was launched in 1863, former New York Senator John Adams Dix was made president and successful Massachusetts railroad construction magnate Thomas Clark Durant vice-president, joined by a lengthy list of northeastern directors designated by Congress. Durant assumed the burdens of management and money raising, and with much money at his disposal he convinced Congress in 1864 to double the size of the Union Pacific's land grant to eighty miles wide and increase its privileges, especially in contracting private debt.[31]

In 1864, Congress also authorized the Northern Pacific to build a railroad from Duluth, Minnesota, to Seattle, Washington, and Portland, Oregon. The Northern Pacific received a grant as generous as the modified Union Pacific award: Twenty odd-numbered alternate sections per mile on each side of the tracks in all territories traversed and ten per mile within states.

Additionally, for all lands within the railroad grant classified by the government as mineral land or unavailable for some other reason such as previous occupation, the railroads might choose "in lieu" lands, also known as "indemnity selections," within ten miles of the outer line of the regular alternate section grant, a swath in some places one-hundred twenty miles wide.[32]

Northeastern core capitalists received the Northern Pacific's land grant. A whole list of them was named by law, but Jay Cooke gained effective control. The financing needed to rebuild the South and to develop the West was ultimately dependent on the cooperation of northeastern core capitalists such as Jay Cooke. Their might was such that they controlled or influenced economics and politics throughout the nation well into the twentieth century.[33] Indeed, the northeastern money trusts of the nineteenth century are active today in only slightly altered form and we shall meet them later in this book.

If we look at the developing West in the period 1865-75 from the perspective of the financier, we see a vast, extremely rich though mostly undeveloped resource base, dependent on eastern financial interests for development capital. It takes money to develop mines, build railroads and cities and expand agriculture. That money ultimately had to come primarily through the hands of the northeastern financial and industrial magnates.[34]

The eastern capitalists could scarcely imagine a more lucrative investment climate. Virtually anything that happened in the West depended on their financial and often political cooperation. They took full advantage of their power. For instance, railroad freight rates for eastern goods shipped west were considerably less than for western goods shipped east. Ironically, that inequity remains.

Only one factor could upset this near-monopoly: the very richness and vastness of the West's resource base was such that if a sufficient portion of that resource wealth fell into hands not controlled by eastern interests, the West could industrialize and become an industrial and financial center in its own right. Such competition was unacceptable.[35]

Northeastern interests worked diligently to influence western land use laws over the next few decades.[36] Keeping large segments of the resource base out of the hands of the individual included preventing the conveyance of fee title on federal range to the ranchers who now held a possessory interest in the federal split estate. Maintaining the one hundred sixty acre maximum in the Homestead Act was vital to their objectives.

The chaos and bitterness ensuing from the Civil War served to maintain among easterners a perceived threat of political instability in the West.[37] As long as that perceived threat lingered, it provided a valid reason for easterners to keep restrictive western land laws in force and prevent the rapid conveyance of title to the West's federal lands into private hands. This was to the advantage of both the General Land Office personnel and the northeastern financial interests. [38] The General Land Office agent kept his fee and salary system intact. As long as the resource-rich lands remained unfettered with private title the opportunity for northeastern interests to profit from new mineral discoveries and other development was greatly enhanced.

As time went on, the financiers were able to dominate these new discoveries without the inconvenience of fighting against settlers and their private property safeguards. It was much simpler and cheaper to buy the mineral rights to a prospector's small mining claim than to buy the whole bundle-of-rights to a segment of a rancher's vast range—a rancher who might view mining as a threat to the viability of his entire ranch.

Until 1875, the unsettled post-Civil War conditions served as one of the excuses for maintaining a huge federal land base in the West.[39] From 1871 forward there were increasing demands for a change in land-use laws to recognize protected private property interests in the federal lands and convey title accordingly.[40]

The northeastern interests—the Rockefellers, Morgans, Goulds, Carnegies and Vanderbilts, among others who had profited handsomely from the Civil War—were loath to share the resources of the South and West equitably with common citizens. The West in particular made a fine colony for the eastern elite. As Walter Dean Burnham said, "The United States was so vast that it had little need of economic colonies abroad; in fact it had two major colonial regions within its own borders, the postbellum South and the West."[41]

By 1890 a Massachusetts Republican, Representative Elijah Morse, could say during a debate with a silver bimetallist:

> During this debate I have heard from gentlemen on the other side deriding words spoken of the businessmen, manufacturers, bankers, and capitalists of New England and Massachusetts. They have been derided in these halls as "bloated bondholders," "gold-bugs," "shylocks," etc.
>
> Neither New England nor Massachusetts needs any eulogium or defense at my hands.... Eliminate from the West and South the capital and enterprise of these same men, and I tell you the unbroken prairie and the primeval forest of creation would stand in place of some of your cities and populous towns. Who builds the Western railroads that develop that great and growing section of the country? Who joined the Atlantic and Pacific Oceans with a railroad? Who builds the lofty buildings, whose summits pierce the sky, in your Western cities?
>
> I answer. . .New England enterprise and New England capital.[42]

But as early as 1870, which was well into the Reconstruction era, northeastern interests became keenly aware that the mountain ranges of the West contained most of the nation's mineral wealth. New gold and silver finds in the

West routinely appeared only in the mountains, for example, the Black Hills of South Dakota (1874) and Cripple Creek, Colorado (1891). These rugged geologic structures were open and exposed to exploration while the intervening valleys were basically blind to mineral exploration and already filling with settlers. The northeastern interests put one item at the top of their political agenda: keep the mountain ranges of the West out of private hands until they could be fully explored for mineral values. The northern core interests feared the already-established private interest in water rights and range rights on the surface would lead to private fee title ownership of the entire land by ordinary citizens.[43]

Of equal if not greater significance was the potential for hydro-electric power generation in the watersheds of the West's mountain ranges. Electricity was the high technology of the last quarter of the nineteenth century. Northeastern financial interests moved to monopolize the best hydro-power sites. These interests often viewed existing water rights and range rights claimed by stockmen as an intolerable obstruction to their objective.

McCarthy and Lamm in their book *The Angry West* comment on the impact of the watershed protection issue in serving the economic objectives of northern core financial interests.

> Throughout most of the century, huge politico-economic combines—many with eastern ties—dominated the critical western water-power industry. By 1914, some twenty-eight corporations owned 90 percent in Utah, and 20 percent in Idaho. Montana Power owned 97 percent of the power development in that state, and as late as 1970, Montanans held only 36 percent of the company's stock.[44]

The watershed protection theme was central to the forest reserve concept. The ostensible purpose of the forest reserves—to protect forests for watershed preservation, aesthetic enjoyment, and game conservation—is variously attributed to 19th century conservationists such as pioneer landscape architect Frederick Law Olmsted, naturalist John Muir, or efficiency-minded forester Bernhard E. Fernow. In fact the idea appears to have originated much earlier in the smoke-filled rooms of 1860s establishment politics.[45]

The Union Pacific and Northern Pacific railroad capitalists convinced the government to withdraw all indemnity lands from settlement pending selection by the railroads. The railroads were in no hurry to choose, which kept settlers out of huge and desirable tracts of land—over twenty-one million acres—that would otherwise have been available to homesteaders.

Twenty years after the Civil War, the policy was still in place. In 1886 a complaint from a settler named Guilford Miller came before the General Land Office because Northern Pacific had claimed his homestead as part of its indemnity selection. Secretary of the Interior Lucius Lamar sent the dispute to the United States Attorney General for a legal opinion. The Attorney General in 1887 upheld the railroads' contention that a settler could acquire no rights or title to public land while the railway privilege of selection within a prescribed area was pending.[46]

The injustice was so rank that President Grover Cleveland wrote an executive order allowing the Interior Secretary to withdraw land from the contingent indemnity lists and reopen it to settlement.[47] Lamar reopened over 21 million acres. This case represents the first documented instance this writer can find in which northeastern capitalists and the government cooperated to use federal reservations as a means to prevent occupancy of potentially valuable lands by settlers.

The second instance involves the creation of Yellowstone National Park. In 1870, Jay Cooke & Company financed a nineteen-man party of Montana dignitaries to explore the myth-shrouded headwaters of Lewis and Clark's *La Roche Jaune* (the River of the Yellow Rock). Mountain men such as Jim Bridger and John Colter in the early 1800s had spread improbable tales of spouting geyser fountains, boiling mudpots, and fairy-ring mineral deposits there—a potentially lucrative attraction for railroad customers by the million, if the stories were true.

Cooke secured an Army escort of one corporal and four privates commanded by Lieutenant Gustavus C. Doane and convinced Henry D. Washburn, the Surveyor-General of Montana, to lead his expedition. Prominent Montana citizens including Judge Cornelius Hedges and Nathaniel Pitt Langford went along.

Langford worked with Jay Cooke and Cooke's management officer A. B. Nettleton. Whatever was to come of the expedition, it must unfailingly accrue to the benefit of the Northern Pacific: Cooke's efforts to raise the hundred million dollars he needed to capitalize his railroad were flagging and Jay Cooke & Company needed a good jolt of enthusiasm to push NP tracks into Western Montana—still five hundred miles beyond the end of steel at the time.

When the Washburn expedition discovered that all the tall tales were true, the party began to develop a method for exploiting their find. Langford wrote in his journal:

> The proposition was made by some member that we utilize the result of our exploration by taking up quarter sections of land at the most prominent points of interest, and a general discussion followed. One member of our party suggested that if there could be secured by preemption a good title to two or three quarter sections of land opposite the lower fall of the Yellowstone and extending down the river along the canyon, they would eventually become a source of great profit to the owners. Another member of the party thought that it would be more desirable to take up a quarter section of land at the Upper Geyser Basin, for the reason that that locality could be more easily reached by tourists and pleasure seekers. A third suggestion was that each member of the party preempt a claim, and in order that no one should have an advantage over the others, the whole should be thrown into a common pool for the benefit of the entire party.

Mr. Hedges then said that he did not approve of any of these plans—that there ought to be no private ownership of any portion of that region, but that the whole of it ought to be set apart as a great National Park, and that each one of us ought to make an effort to have this accomplished. His suggestion met with an instantaneous and favorable response from all—except one—of the members of our party, and each hour since the matter was first broached, our enthusiasm has increased.[48]

The effect was to monopolize a major tourist attraction for an eastern transportation capitalist and prevent exploitation by competing westerners. The expeditioners went from Yellowstone to Washington. With the leadership of Langford and the money of Cooke they spent the winter of 1871-72 lobbying Congress to create Yellowstone National Park. The rhetoric was all nature appreciation and public enjoyment. The real purpose was monopoly for the Northern Pacific. After listening to testimony assuring skeptical members that the proposed forty square mile park was worthless for settlement, mining or timber, on March 1, 1872, Congress obliged.

Jay Cooke and the Northern Pacific gained exclusive concession rights to serve Yellowstone National Park. Nathaniel Pitt Langford became the first Superintendent of Yellowstone National Park. Federal troops helped him manage Yellowstone. Yellowstone National Park was the invention of commercial interests, lobbied by commercial interests, monopolized for commercial interests.[49]

THE RESERVE CONCEPT

Once the device of reservations was well established for park purposes, the idea spread to general forest reserves. The first legislation introduced in Congress proposing the withdrawal of public land from settlement for forest reserves was introduced in 1876 by Greenbury L. Fort, congressman from Illinois. No record survives of where Fort got the idea of legislating forest reserves by removing public land from settlement, but it may have been inspired by Dr. John A. Warder, founder of the first American Forestry Association. Warder was acquainted with Fort and was also well aware of an 1873 paper titled "On the Duty of Governments in the Preservation of Forests" by Franklin B. Hough, a medical doctor, amateur historian, and botanist from Lowville, New York. Hough himself probably got the idea of forest preservation from George Perkins Marsh, a New England judge whose 1864 book *Man and Nature* qualifies him as America's first political ecologist.[50]

Nonetheless, it's equally possible that Fort got the forest reserve idea from the Union Pacific and Northern Pacific capitalists. Their social set was widely read and well aware of Judge Marsh's *Man and Nature*. They kept abreast of all resource-development news, including Dr. Hough's paper on forest preservation. With their political acumen, it is inconceivable that they did not extend the railroad indemnity reserve and national park reserve concepts to the withdrawal of public lands from settlement as forest reserves—or that they failed to realize that the growing conservation movement could be used to this end.

However, a strong political reaction to large northeast trusts was building in America. Fort's 1876 bill failed, as did nearly two hundred separate forest reserve bills until 1891, when almost by accident the concept became law. While these bills were failing year by year, eastern capitalists were quietly dominating the western range.

THE CATTLE CAPITALISTS

Among the many excesses of the Reconstruction era, 1865-77, was the plunder of the western range by eastern capitalists and their investment cliques. Historian Gene M. Gressley noted that, "With little distortion, it could be said that the membership of the Union League Club of Chicago, the Union League Club of New York, and the Harvard Club of Boston controlled the major portion of Eastern capital in the range cattle industry."[51]

The next most dominant investor clique was that of the railroad magnates. Union Pacific executives Sidney Dillon, Thomas C. Durant, and Oliver Ames together established an immense ranch near North Platte, Nebraska, in the 1880s. Railroad genius Sidney Dillon served on the board of directors of the Union Pacific from 1865 to his death in 1892—and the town of Dillon, Montana, is named after him. The Kansas City *Livestock Indicator*, on October 16, 1884, printed:

> Sidney Dillon, noted ex-potentate of the Union Pacific is one of the heaviest cattle owners in the West. The firm of which he is principal is known as Dillon & Co. Isaac Dillon is in the company and as a manager of the business, lives on a ranch near North Platte. The cattle are numbered by the thousands on a thousand hills. Some in the Platte Valley, others on Snake River country. The headquarters known as Stone Ranch, is in the valley near O'Fallon station and surrounded by a splendid range.

Railroad executives in general capitalized on their distinct advantage and captured huge slices of the western domain. General Jack Casement, a construction contractor for the Union Pacific, bought up land in Kansas, Nebraska, and Colorado and soon had herds on his private lands. John B. Alley, another Union Pacific director, usurped the presidency of the Palo Blanco Ranch Company in New Mexico after its former leader, one Stephen Dorsey, became tainted in the 1881 Star Route postal frauds.[52]

Jay Cooke had accumulated heavy landholdings in the upper Midwest, which were managed for his estate in bankruptcy by J. C. Young, a Cedar Rapids, Iowa, attorney, and J. Horace Harding, trustee of the estate. Cattle made up an important part of Cooke's rental income.

Historian Gressley identified 93 eastern cattle and land speculators who invested more than $10,000 into the boom era. The list includes such well known names such as Vanderbilt, Whitney, Roosevelt, Rockefeller, Goodrich, Busch, and Case. Some lesser known individuals such as George Bird Grinnell, Clarence King, Alexander Agassiz, and Preston B. Plumb were also on the list and would later play key roles in the establishment of the forest reserves.[53]

It is useful to briefly acquaint ourselves with some of these men who would so profoundly affect the western range problem of today.

New York: John and Charles Arbuckle were coffee manufacturers who invested in the PO Ranch near Cheyenne, Wyoming, and in the Lea Cattle Company of Roswell, New Mexico. Robert Bacon was a partner in the House of Morgan, the same J. P. Morgan who ruined Jay Cooke and precipitated the Panic of 1873; Bacon owned stock in the Riverside Land and Cattle Company and the Teschemacher and deBillier Cattle Company, which owned five ranches in Montana and Wyoming. James Gordon Bennett, Jr., published the *New York Herald*, the sensationalist paper that financed Henry Stanley's expedition to Africa to find David Livingston; Bennett owned ranches in Nebraska and Wyoming.

John Bigelow, author, diplomat and founder of the New York Public Library, was a friend of the House of Morgan, and joined Robert Bacon in the Teschemacher and deBillier Cattle Company. Boston's Hubert E. Teschemacher and New York's Frederick O. deBillier met as students at Harvard and obtained financing for a western cattle ranch from Boston bankers Joseph Ames Senior and Junior.

William Andrews Clark was a Pennsylvania native who made a huge fortune in Montana copper mines, bribing his way into power through corrupt Butte Judge William Clancy who shut down rival mines for a substantial cash consideration. Clark overcame competitors by tying the exploitation of copper to eastern capital, using his partnership with John D. Rockefeller's Standard Oil associate Henry H. Rogers to capitalize the Amalgamated Copper Company at seventy-five million dollars. No dollar ever got away from Clark, it was said, without coming back stuck to another. He was fraudulently elected U. S. Senator from Montana and resigned after a congressional investigation but later won fairly and served a full term. Clark's ownership of large Montana cattle ranch holdings was almost an afterthought, just something else to do with eastern capital.

Rufus Hatch, Wall Street stockbroker, confined most of his ranching activities to brokerage in livestock companies and lands, particularly during the development of the Chicago and Northwestern Railroad. Abram S. Hewitt was an ironmonger and the son-in-law of Peter Cooper, who was one of our 1855 millionaires who made his fortune manufacturing glue and isinglass. Hewitt joined fellow New Yorker James Gordon Bennett, Jr., in ownership of ranches in Nebraska and Wyoming.

Augustus Kountze, a banker, had immense holdings in western cattle, real estate, lumber, and railroads. The Kountze Brothers firm, including Augustus, Herman, Luther, and Charles B., had banks in Denver, Indianapolis, Omaha, and New York. Their richest cattle payoff came from an association with flamboyant cattleman of the Texas Gulf Coast, Able Head "Shanghai" Pierce, which investment encompassed 200,000 acres and 12,000 yearlings.[54]

William Rockefeller, younger brother of John Davison Rockefeller, needs no introduction except as a cattle baron. Although the Rockefellers owned western holdings that are legion, the public knows them best for forming the

Copper Trust that swallowed the Anaconda and Amalgamated of Montana, and for their war with the Guggenheim copper kings to get them to sell out to the Rockefeller smelter trust called the American Smelting & Refining Company (ASARCO). Their cattle holdings were extensive.

Theodore Roosevelt, when he is thought of as the son of a wealthy eastern glass merchant rather than as the Trust-Buster preaching from his "bully pulpit" of the presidency, becomes an obvious sectionalist. Roosevelt bought Dakota Badlands ranches at the advice of A. D. Ferris and tried to influence the Stockmen's Association.[55]

Dick Trimble was sent West by his father to help his Harvard classmates Teschemacher and deBillier manage their ranch. Trimble is important because his story is symptomatic of a whole movement during the 1880s. Many an eastern capitalist confronted with an unruly young son or a scion involved in some scandal had the offender "sent West" in hopes of reforming and rehabilitating the wayward offspring. The magnates deemed a job "managing" a ranch extremely beneficial. Most frequently, however, the "exiled" city dweller quickly discovered that Cheyenne and Denver offered—as historian Gressley described it—lush and lusty clubs with long bars where wealthy cattlemen gathered to forget the problems of the range in a glass of good bourbon or a special stage presentation. In actual practice, these "resident absentee owners" managed nothing except to delay and exasperate the foreman, who tried to run the spread despite the eastern "help." Trimble, incidentally, reformed and joined the House of Morgan to become the first secretary-treasurer of U. S. Steel.

Cornelius Vanderbilt, shortly after the New York Central built extensive cattle yards in Kansas City during the 1870s, directed his son William H. and grandson William K. Vanderbilt to organize a huge livestock company in Hermosilla, Colorado. The Vanderbilts liquidated their "charter" stock in the firm in 1879 at a fine profit.[56]

Perhaps more significant than all these for the future of United States public land policy was George Bird Grinnell, an anthropologist and ethnologist who later became publisher of the influential periodical *Forest and Stream*, which promoted federal intervention in the form of forest reserves. Grinnell is credited by some as the inventor of the concept of designated wilderness areas. He preempted a ranch near Lander, Wyoming and later helped promote the creation of forest reserves.

Boston: Alexander Agassiz was the son of imminent geologist Louis Agassiz and, along with his brother-in-law Quincy A. Shaw—a copper magnate of vast means—provided fellow Bostonian N. R. Davis with capital to start the N. R. Davis Cattle Company in Wyoming and Nebraska. Agassiz also lent Thomas Sturgis, secretary of the Wyoming Stock Growers Association and president of the Union Cattle Company of Cheyenne, a substantial amount for expansion. Agassiz also invested in the Union Pacific mines at Telluride, Colorado, farms in Colorado and real estate in California.

George Blanchard, mercantilist, bought shares in the North American Cattle Company, sold them, and then bought a seat on the board of directors of the Western Union Beef Company, a chair he held for eight years.

The Converses, James and Edmund, were dry-goods merchants who, along with friend Augustus Gurnee, invested in the Teschemacher and deBillier Cattle Company. Edmund, with fellow Bostonian B. W. Crowinshield, also lobbied legislators to pass the Animal Industry Bill bringing federal regulation to the livestock industry. George Denny and Henry A. Rice, partners in a brokerage house, directed capital to many western livestock companies.

John Haldane Flagler, whose first wife was a member of the Converse family, held a seat on the board of the Ames Cattle Company of Beaverhead County, Montana, and invested heavily in Utah copper, Kansas Pacific, Union Pacific, and Dolores Mining stock.

Chicago: Financier Henry Blair and partner Richard Lake funnelled huge eastern core investments to herds running from the Dakotas to Texas. John V. Clark was only nineteen when he assumed a dominant role in the Porter Land and Livestock Company of Colorado. John Clay, prominent banker and commission man, took over from Godfrey Snydacker (First National Bank of Chicago) the job of untangling the books of the insolvent Swan Land and Cattle Company. Clay controlled Western Ranches, Ltd. in South Dakota, while in another office of Chicago's Rookery Building, Joseph Rosenbaum controlled several cattle corporations in Montana.

R. M. Fair and department store partner Marshall Field were convinced by Chicago real estate magnate Levi Leiter to become directors of the enormous Pratt-Ferris Cattle Company with holdings in Nebraska, Wyoming, and South Dakota. Nelson Morris, meatpacker and member of the commission firm of Gregory Cooley and Company, accumulated ranches in Texas, Wyoming, and Montana. He never even saw his immense 250,000 acre Texas holdings. Morris became an influential member of the American Cattle Trust, and lobbied against the Animal Industry Bill. It is worth noting that Senator Preston B. Plumb of Kansas, who helped create the Animal Industry Bureau in the federal government to regulate the livestock industry also held western cattle investments through a Chicago brokerage house, according to historian Gressley. Plumb was later instrumental in guiding the 1891 bill that authorized the forest reserves that became our present national forests.

Portus B. Weare, a wealthy commission man, organized the Weare Land and Livestock Company in 1877, that four years later had 50,000 head of cattle grazing on the public domain in Montana, Wyoming, and South Dakota.

Not far from Chicago, in DeKalb, barbed wire tycoons Joseph Glidden, Isaac Ellwood, Henry Sanborn and John Warne Gates ran a vast empire of cattle holdings all over the West. Sanborn founded Amarillo, Texas, Ellwood surrounded Lubbock with three ranches totalling 246,347 acres costing over two million dollars, Glidden invented barbed wire, and Gates sold it extensively in the West.

Colorful John "Bet-A-Million" Gates is remembered by historian Stewart Holbrook for hawking barbed wire fencing as being "Light as air, stronger than whiskey, and cheaper than dirt." It was Gates who put together the barbed-wire trust, which in 1898 was folded into the American Steel and Wire Company, which in its turn became part of the world's first billion dollar trust, U. S. Steel. Gates was the man who helped J. P. Morgan buy out steel magnate Andrew Carnegie so that U. S. Steel could be formed, and initiated the negotiations to buy out Union Steel, Andrew Mellon's nuisance company, according to historian Holbrook, "at a fantastic thirty million dollars, much more than it was worth."[57]

Of the eastern investors in western cattle outside the Boston-New York-Chicago triangle, some need no introduction: David M. Goodrich was a scion of the B. F. Goodrich family and its rubber fortune. Others less well known were no less remarkable, such as Washington D. C.'s Clarence King, eminent geologist for the War Department, who served briefly as the first director of the U. S. Geological Survey, and who invested in the N. R. Davis Cattle Company and the Dakota Cattle Company. He received returns as high as fifty percent on his Davis investment.

These men were to have a profound influence on the West. Their failed attempt to gain ownership of the West for themselves in the latter half of the nineteenth century was a major factor leading to today's storm over rangelands. When I refer to "northeastern core interests" or "eastern capitalists," I am primarily referring to these men and their close associates.

NOTES ON CONSERVATION
Conservation as a basic attitude toward natural resources has been around since man's beginnings. The earliest pots and containers shaped by prehistoric humanity were intended to conserve resources. Every famine in the history of mankind has spurred the more intelligent to realize that wasting natural resources is not a wise practice. In the seventeenth century, Thomas Robert Malthus dramatized the idea with his predictions of universal doom based upon projected population growth always outstripping available food supply.

The Civil War consumed immense quantities of natural resources in a highly visible fashion. For example, extensive timber cutting was necessary to provide charcoal for gunpowder manufacture. We should not be surprised then to see concern over natural resource utilization emerge in the United States in the 1860s.

When George Perkins Marsh wrote *Man and Nature* in 1864, he was complaining in a large degree about the wholesale removal of the white pine and hardwood forests that characterized the Atlantic seaboard states. When another amateur naturalist, Reverend Frederick Starr, raised the alarm of a wood famine in 1865, he was reacting to the extensive clearcutting of slow-growing hardwood forests in the East.

When historians Frederick Jackson Turner, Frances Parkman, and even writer Rudyard Kipling later joined in the conservation chorus, their focus was primarily on the extensive utilization of forest resources east of the Mississippi.[58]

By the late 1870s the conservation movement was gaining social acceptance in influential circles. It was also becoming troublesome to northeastern industrial and financial interests who were profiting from the harvest of eastern forests to rebuild northeastern cities after the Civil War.[59]

A series of events came together in the late 1880s and early 1890s that was to dramatically affect land policy in the United States. The nation had slumped into an economic depression.[60] As in earlier times of national financial stress, the Atlantic states found a convenient whipping-boy for their difficulties in the "cheap and excessive production" of a developing West.[61] The lumbermen in the Northeast saw the vast timber resources of the West as a cause of their immediate problems and a threat to their future. Eastern farmers saw the developing crop land of the West as a reason for their depressed produce market. Eastern stockmen, who were forced to sell their animals below cost in a depressed meat market, blamed cheap production from those "free rangelands" in the West. The miners in the eastern states who had closed their operations because of low prices for their output saw the threat of a mineral-rich West as their ultimate demise. These people voted and politicians responded to their fears and concerns, then as they do today. The pervasive East-versus-West animosity was aggravated considerably by the bad economic conditions of the era.

The political and economic climate was right during the late 1880s for eastern financial and industrial interests to focus a troublesome conservation movement on the developing West. If the nature-protection philosophy of the conservationist movement could be turned into an obstructionist lid on western economic growth and production, it could only help depressed eastern markets. Eastern industrial support for the conservation of resource lands in the West began to quietly spread under the economic exigencies of the late 1880s.

PREEMPTIVE RANGE RIGHTS

While forest reserve bills were failing year after year in the final quarter of the 19th century, one overriding concern emerged among northeastern interests. This concern was the elaboration of preemptive property rights by ranchers attempting to define their status in the vast rangeland areas.[62] To prevent overgrazing and damage to the resource base, ranchers needed definite boundaries for their operations, with fences to control livestock use.[63] They realized that some definition had to be instituted to prevent the "Tragedy of The Commons" from becoming the epitaph of the western rangelands.[64]

Ranchers gained title to the use of stock water on the federal range from the individual states. Proof of beneficial use to satisfy the demands of local law and customs and rules of the court was provided by demonstrating that the rancher had a legal right to graze. That legal right was based on the time-honored legal doctrine of "first in time, first in right."*

* The prior appropriation doctrine is explained in full at pp. 135ff.

Ranchers began building fences to protect their claimed rangeland from overgrazing by livestock operators who could not demonstrate a legal right to graze according to local law and custom and rules of the courts. In other words, western ranchers were laying claim to vast segments of the West by preemption, a settlement process widely practiced in the eastern portions of the country.

From 1830 through 1890 Congress had recognized preemption as a legal method of land settlement on the nation's frontier.* Congress had passed numerous acts accommodating the practice usually limiting the preempted acreage from forty to one hundred sixty acres. Forty to one hundred sixty acres was considered an economic unit east of the Mississippi.

Until Congress had passed the preemption laws, settlers on the public domain were technically trespassers. A property right by use, or in other words, a prescriptive right or squatter's right—while often recognized by local law and custom and court decisions in settling local disputes—was not good against the federal government.

It is a well-settled matter of law that prescriptive rights cannot be asserted against the sovereign. However, if the sovereign determines to recognize prescriptive rights, it may do so by granting its permission through laws such as the preemption statutes of the 1830-1890 era.

A government grant of permission makes possible the conversion of the locally recognized prescriptive property right to a federally recognized preemptive property right. Recognition of the preemptive right by the sovereign can lead to the patent of the land in fee simple title.

Western ranchers were following the same principles of preemption their eastern counterparts had used. The major difference was the absence of a law that would define an economic unit in the West.[65]

Obviously, one hundred sixty acres was not an economic unit on the West's arid rangelands where many sections of land were required to support a viable ranching operation. Let's look at some of the problems this presented.

These rangeland claims covered most of the vast timberlands of the West with their untold wealth of timber resources. These claims covered virtually all of the mineral-rich mountain ranges of the West. If western interests were allowed to lay claim to the vast resource bounty through preemption, an entirely new class of financiers and industrialists would no doubt emerge in the western states. Political power would follow. Northern core interests were justifiably alarmed.[66]

The Interior Department, particularly President Hayes' Interior Secretary Carl Schurz (served 1877-81), responded to the threat of preemptive property rights by vigorously opposing fenced boundaries. A federal law in 1885 forced removal of fences built to protect range from overgrazing, sometimes with federal troops standing by to enforce the measure. Throughout the 1880s efforts to annul preemptive range rights were intense.[67]

WATER CONFLICTS

The first settlers in the arid west had by necessity laid claim to the water, the essential substance upon which settlement was dependent. Sometimes, particu-

*Supra, p. 43.

larly along streams and rivers, settlers claimed both land and water. By this action they could control adjacent federal range lands as long as no other source of livestock water was present.

Where scattered springs, seeps, and wells were the stockwatering sources, settlers quickly laid claim to these under the prior appropriation doctrine. Control of these scattered water sources gave the claimant control of sometimes vast acreages of land for grazing purposes. It was commonly acknowledged in the arid West that he who owns the water owns the land.

The race was on to own the water and by extension the land. Settlers were viewing the general grant of permission expressed toward agriculture in the Act of July 26, 1866 as the legal vehicle through which their prescriptive range rights would be viewed by the sovereign as preemptive range rights subject to conveyance of title from the federal government.

Eastern and foreign capitalists also saw the opportunity to acquire title to huge tracts of western land by the same methods. Preemptive range rights could convey title not only to the range, but also to the timber, minerals, hydro power sites, etc., embodied in the land. The tremendous competition for control of the western ranges, characterized by the historically popularized range wars and overgrazing, was a competition for ownership of the land and its wealth of resources. Livestock grazing was a means to that end. Eastern and foreign capitalists moved vast herds of livestock into the West to assert their own claims.

The eastern or foreign capitalist often viewed the early farmers and stockmen as barriers to their ability to establish preemptive rights to the West's land base. The early stockmen and farmers, by virtue of the doctrine of prior appropriation, often had legal control of the water and the land. If the later arriving capitalists were to achieve their goal of gaining control of the land they first had to disenfranchise those with prior claims.

Large bands of sheep were often used to accomplish the objective of "eating out" the small stock grazer or homesteader. A common ploy was to graze bands of sheep near the private land or waters of the small settlers until the small settler was forced to abandon his land and water claims. By constantly overgrazing the acres adjacent to the land and water claims of the small settlers, large corporate livestock organizations hoped to disenfranchise the original prior appropriator and gain a preemptive position on the resource base for themselves. Sometimes cattle herds were used in the same way. In other instances the large companies were able to gain control of the water by these means and force the small livestock man to beg for access to water for his stock.

Congressman William McAdoo of New Jersey made some revealing statements about the difficulties faced in attempting to gain control of land and water of prior appropriators on the rangelands of the Northwest. The occasion was a debate that took place on June 26, 1888, over an amendment to a major new land law bill, H.R. 7901. The issue was the availability of stock water in the arid West and the control of that water by prior appropriators. Mr. McAdoo said:

Some two or three years ago it was my good fortune to pass over the Northern Pacific Railroad, and while so doing I had a conversation with one of the largest capitalists of New York City, as well as the owner of the greatest cattle herd in the whole of the western region.

I studied the question while in that country as best I could and the portion of the case as far as the disposal of the public lands was concerned. From personal observation I became convinced that the man who owns the water owned the land virtually. . . .

I think that in my opinion the United States ought, under the right of eminent domain, if no other way, to condemn the right of way for the smallest cattle-owner to take his herd of cattle to the moving streams of water; and this gentleman admitted that in his opinion that was the solution to the whole problem—getting the small cattle herds in such a position that they might afford to live as against the tremendous companies that control the cattle trade of that region.

It is injustice to a man. . .to find that. . .his neighbor. . .owns the river, or the creek, or the lake, or the spring, or that it belongs to the tremendous cattle syndicates, and who hold it in such a tight grip in the hands of these powerful monopolies that a thirsty dog traveling the prairie dare not stop to lap the flowing water to slake his thirst without paying tribute if he could to the powerful combination. . . .

Confederation, combination, and conspiracy, inspired by greed and corporate and individual rapacity, threaten the welfare and very existence of our people.[68]

Throughout this period, northern core policy was a reaction to state control of water arising on the federal lands. The northern core wanted control of that water, but prior appropriation rights of the original settlers kept it out of their hands. Courts upheld the prior rights of early settlers against the eastern livestock interests. Where this happened, northern core capitalists suggested leasing the rangelands as a way to thwart western control of water rights. However, even the lease idea failed because courts protected prior appropriation rights. One truth became ever more apparent: Who controlled the water controlled the land. And the eastern plutocrat intended to control the land.

PLUTOCRATS ON THE FEDERAL LANDS

The proof that these eastern capitalists sought to eradicate prior appropriation rights in the West is provided in the actions of a particular investment clique, one of those "syndicates" McAdoo complained of: the American Cattle Trust. On February 9th, 1887, Edward M. McGillin, a dry-goods entrepreneur of Cleveland, Ohio—and like his neighbors the Rockefellers, a deep investor in the range cattle industry—addressed the International Range Association meeting in Denver, Colorado. His message was simple: the Chicago meat-packers held a monopoly of the cattle market. Monopoly must be fought with monopoly.[69]

McGillin presented to the Range Association his plan for a huge trust, incorporated at one hundred million dollars. The trust would "arrange, manage, sell every animal from the time it was dropped a calf until it was beef in the consumer's basket." The capital was to be raised from subscribing cattlemen who would mortgage their herds at $5 per head, payable in five annual installments subject to 6 percent interest. Payments would be deducted from beef sales by the trust. Trust members could not sell outside the pool at any price. Each state was to be responsible for electing its own directors, and the state presidents would constitute the directors of the national company.

On May 3, 1887, the New York *Times* reported the formation of the Trust.[70] Historian Gene Gressley described the circumstances:

> In organization, the American Cattle Trust was modeled directly after the Standard Oil Trust. In early April, F. K. Sturgis bragged to Henry L. Higginson: "We have (through influence) obtained a copy of the Standard Oil Trust, a most difficult thing to do." Further, Sturgis trumpeted that they had managed to secure the cooperation of the huge Chicago packer Nelson Morris. "It will be the salvation of the ranch interests when I tell you that Morris kills 1,000 steers daily and utilizes them in every part; you will readily see the gain will be very great if the ranch interests can share the slaughtering profits." Sturgis closed by predicting that soon the Cattle Trust certificates would be listed on the stock exchange: "I see no reason why they should not be as good, or better, than the cotton oil trust certificates.[71]

The Trust developed rapidly. Morris's Chicago plant was purchased for two million dollars. Feeding farms at Gilmore, Nebraska, next to the Union Cattle Company, were purchased. Contracts for canned beef were signed with the French and Belgian governments. A tough roster of managers was hired and went to work. Western cattle companies applied in droves for admission to the Trust.

Conceived with the eastern capitalist mentality, the Trust went contrary to everything that had built the western range cattle industry in the first place: it parceled out herds to its network of skilled managers without regard to water rights, range condition, competing livestock interests, settlers, or range rights. This astounding fact is documented beyond dispute in the Trust's surviving letterbooks and is remarked upon by historian Gressley.[72]

As soon as the eastern capitalists got a substantially concentrated hold on the western cattle industry, they set about to abolish the water rights and range rights of their own Trust members. It was no error that the trustees allocated herds without regard for water rights or range rights: trustee orders to local managers preserved in the Trust's letterbooks are clear and calculated, as are the complaints of local managers against the abuses. The eastern capitalists of the Trust knew what they were doing.

However, the plan did not work. Soon every level of management was disagreeing with every other level, and the managers with the capitalists. The eastern trustees argued with the western trustees. By the end of the first year, the spring of 1888, the great monopoly owned 218,934 cattle plus land, equipment, and the packing plant, total value, $7,959,071.61. Considering that the number of cattle outside the Trust in Wyoming alone was 753,648 valued at over ten million dollars, it had failed to achieve much of a monopoly. It was simply one set of eastern interests working to swallow others, and the others were bigger. The squabbling Trust managers couldn't compete with the non-Trust ranches that remembered whose water rights were whose and what range rights meant. The Trust was bankrupt by 1889 and liquidated in 1890.

The American Cattle Trust showed the eastern capitalists that no single investor clique could dominate the decentralized industry. Conditions were simply not mature enough: the industrial infrastructure had not been adequately developed in the West to support a trust of the Standard Oil type. The eastern capitalists realized that economic means wouldn't work, so they switched to political means. Only by invalidating the water and range rights of the individual rancher could wholesale domination be obtained. The political direction was clear, based upon experience with railroad indemnity lands and the success of the Yellowstone National Park reserve.* If the resources of the West could be politically locked up in some kind of reserves until the East found a practical way to exploit them, the objective of dominating the western states could be realized. And that required the end of privatization and the extinction of water rights and range rights.

BLOCKING PREEMPTIVE RIGHTS

Probably the most effective instrument at blocking privatization of the public domain on range, mineral and timberlands alike was the authorization of forest reserves in 1891.

It is doubtful if the forest reserve clause, Section 24 of the 1891 General Land Law Revision Act, would have emerged had it not been for one important catalyst: the United States Supreme Court ruling in the case of *Buford* v. *Houtz*, February 3, 1890.[73]

The court held:

> We are of opinion that there is an implied licence, growing out of the custom of nearly a hundred years that the public lands of the United States, especially those in which the native grasses are adapted to the growth and fattening of domestic animals, shall be free to the people who seek to use them where they are left open and unenclosed and no act of government forbids this use.... The government of the United States, in all its branches, has known of this use, has never forbidden it, nor taken any steps to arrest it. No doubt it may be safely stated that this has been done with the consent of all branches of the government and, as we shall attempt to show, with its direct encouragement.[74]

Buford v. *Houtz* involved the classic range issues of private water rights and range rights on federal lands. The court discussed the case in light of the legal doctrines of prior appropriation and preemption.

Buford was a cattleman and large land owner. Houtz was a sheepman who

*Supra, p. 59-61.

owned no land, but grazed his sheep on the federal lands. Buford argued that Houtz had no "right title or possession, of or to any of the land." Buford asserted that Houtz was nothing more than a trespasser on the federal lands. Buford's argument was the embodiment of the *northern core's* position on preemptive range rights. Houtz's position was the embodiment of the *western states'* position on preemptive range rights.

The court said that Congress had granted its permission for Houtz to graze on the federal lands by allowing and encouraging grazing on the federal lands for one hundred twenty years. The court discussed this liberality of Congress toward livestock grazing on its lands as an integral part of the preemption method of settlement.

> The whole system of control of the public lands of the United States as it has been conducted by the government, under Acts of Congress, shows a liberality in regard to their use which has been uniform and remarkable. They have always been open to sale at very cheap prices. Laws have been enacted authorizing persons to settle upon them, and to cultivate them, before they acquire any title to them. While in the incipiency of the settlement of these lands, by persons entering upon them, the permission to do so was a tacit one, the exercise of this permission became so important that Congress by a system of laws called the Preemption Laws, recognized this right so far as to confer a priority of the right of purchase on the persons who settled upon and cultivated any part of this public domain. During the time the settler was perfecting his title. . . both he and all other persons who desired to do so had full liberty to graze their stock upon the grasses of the prairies and upon other nutritious substances found upon the soil.[75]

The court justified Houtz's grazing rights by comparing them with federal recognition of the private property rights miners acquired under the claim of discovery.

> . . .and while this remained for many years as a right resting upon the tacit assent of government, the principle has been since incorporated into the positive legislation of Congress, and to-day the larger part of the valuable mines of the United States are held by individuals under the claim of discovery. . . .
>
> But this court, reverting to the liberality of the government in that regard, decided that the moment the ore became detached from the main vein in which it was embedded in the mine, it became the property of the miner, the United States having no interest in it. . . .[76]

The 1872 Mining Law further allowed for the patent of the mining claim in full fee simple as private property.

With these clear references to the principles of prior appropriation and preemption in the text of the court's opinion, it is not surprising that many believed preemptive range rights—created through prior appropriation—were now a fact no longer subject to dispute. Although the record is silent, *Buford v. Houtz* must have sent a wave of panic through the anti-western-growth movement. Preemptive range rights and the prior appropriation water doctrine quickly became the focal point of attack expressed in northeastern legislative, litigative, and public relations efforts. Preemptive range rights and the prior appropriation water doctrine continue to be the focal point of attack on the West one hundred years later.

Chapter Three Footnotes

1. Thomas L. Livermore, *Numbers and Losses in the Civil War, 1861-1865*, Houghton Mifflin & Co., Boston, 1900, reprint Morningside, Dayton, 1986.
2. The best modern scholarship covering the entire Civil War era is J. G. Randall and David Donald, *The Civil War and Reconstruction*, 2nd ed., revised, Heath, Lexington, 1969.
3. See generally Ralph Paul Bieber, ed., *Southern Trails to California in 1849*, The Arthur H. Clark Company, Glendale, 1937. For collected letters, journals and other Southern gold rush memorabilia, see also Hezekiah John Crumpton, *The Adventures of Two Alabama Boys*, (self-published), Montgomery, 1912; and Isaac H. Duval, *Texas Argonauts: Isaac H. Duval and the California Gold Rush*, Book Club of California, San Francisco, 1987.
4. Hubert Howe Bancroft, *History of Utah, 1540-1887*, The History Company, San Francisco, 1890, pp. 440-441.
5. Ibid, p. 446.
6. *History of Nevada*, Thompson & West, Oakland, 1881; reprinted by Howell-North, Berkeley, 1958, p. 58.
7. James G. Scrugham, ed., *Nevada: A Narrative of the Conquest of a Frontier Land*, 3 vol., American Historical Society, Chicago and New York, 1935, p. 198.
8. Ibid.
9. *History of Nevada*, Thompson & West, p. 63.
10. James M. McPherson, *Ordeal By Fire: The Civil War and Reconstruction*, Alfred A. Knopf, New York, 1982, p. 72.

11. Nevada Laws, 1864, 81-5. The fact that Nevada was attempting to hold the Knights of the Golden Circle in check by this law is attested in *The Works of Hubert Howe Bancroft, Vol. XXV, History of Nevada, Colorado, and Wyoming, 1540-1888*, The History Company, San Francisco, 1890, p. 183.

12. Effie Mona Mack, *Nevada*, Arthur H. Clark Company, Glendale, California, 1935, p. 187. George Rothwell Brown, ed., *Reminiscences of William M. Stewart of Nevada*, Neale Publishing Company, New York, 1908, p. 129.

13. *History of Nevada*, Thompson & West, p. 84.

14. See Chapter X, "Suburban Nevada," Milton H. Shutes, *Lincoln and California*, Stanford University Press, Stanford, 1943, p. 173ff.

15. Scrugham, ed., *Nevada: A Narrative of the Conquest of a Frontier Land*, vol. 1, p. 198.

16. Walter Van Tilberg Clark, ed., *The Journals of Alfred Doten, 1849-1903*, vol. 1, University of Nevada Press, Reno, 1973, p. 590.

17. Orville J. Victor, *The History, Civil, Political and Military, of the Southern Rebellion, From its Incipient Stages to its Close*, James D. Torrey, New York, 1861, pp. 133-136. Walker's own account of these events is absorbing. See William Walker, *The War in Nicaragua*, S. H. Goetzel, Mobile, 1860; reprinted B. Ethridge Books, Detroit, 1971.

18. Daniel Feller, *The Public Lands in Jacksonian Politics*, p. 44. Although Feller was referring to Ohio in 1802, the characterization is appropriate to Nevada in the 1860s.

19. Hibbard, *A History of the Public Land Policies*, pp. 347-385.

20. *U.S. Statutes at Large*, XII, p. 392.

21. For a thorough explanation of early water rights practices, see Hutchins, *Water Rights Laws in the Nineteen Western States*, vol. 1.

22. Brown, ed., *Reminiscences of William M. Stewart of Nevada*, p. 173.

23. *U.S. Statutes at Large*, 39th Congress, Sess. I, Chap. 262, "An Act granting the Right of Way to Ditch and Canal Owners over the Public Lands, and for other Purposes," pp. 251-253.

24. Gary Libecap, *Locking Up the Range: Federal Land Controls and Grazing*, Pacific Institute for Public Policy Research, San Francisco, 1981, pp. 15-16.

25. Hibbard, *A History of the Public Land Policies*, pp. 424-26.

26. Libecap, *Locking Up the Range: Federal Land Controls and Grazing*, pp. 9-10.

27. The thesis that the capitalist leisure class promoted nature preservation for its own reasons is explored in William Tucker, *Progress and Privilege: America in the Age of Environmentalism*, Doubleday, New York, 1982.

28. An entertaining sketch of America's wealthiest citizens of 1855 is in Stewart H. Holbrook, *The Age of the Moguls*, Doubleday & Company, Inc., Garden City, N.Y., 1956, p. 8ff.

29. Holbrook, *The Age of the Moguls*, p. 11.

30. *U.S. Statutes at Large*, XII, p. 492 (authorization), p. 356 (appropriation).

31. *U.S. Statutes at Large*, XIII, p. 365.

32. "Indemnity" or "lieu" lands are explained in detail in Hibbard, *A History of the Public Land Policies*, pp. 245-249.

33. Henrietta Larson, *Jay Cooke*, Harvard University Press, Cambridge, 1936.

34. See generally William Augustus Croffut, *The Vanderbilts and the Story of Their*

Fortune, Belford, Clarke, Chicago, 1886; reprinted by Arno Press, New York, 1975; Maury Klein, *The Life and Legend of Jay Gould*, Johns Hopkins University Press, Baltimore, 1986; Kenneth D. Ackerman, *The Gold Ring: Jim Fisk, Jay Gould, and Black Friday*, Dodd, Mead, New York, 1988; and Jay Boyd, *The Crédit Mobilier of America*, C. W. Calkins, Boston, 1880, reprinted 1969.

35. The repugnance for competition felt by the eastern capitalists is well illustrated in Holbrook, *The Age of the Moguls*. See particularly the chapters on Rockefeller and J. P. Morgan, pp. 61ff. and 144ff.

36. The details are in Gene M. Gressley, *Bankers and Cattlemen*, Alfred A. Knopf, New York, 1966.

37. Bensel, *Sectionalism and American Political Development: 1880-1980*, p. 73.

38. For details of post-Civil War hegemony enforced through pensions and the tariff, see Bensel, *Sectionalism and American Political Development: 1880-1980*, pp. 62ff.; for General Land Office obstructionism see Libecap, *Locking Up the Range: Federal Land Controls and Grazing*, pp. 7-30.

39. Libecap, *Locking Up the Range: Federal Land Controls and Grazing*, pp. 1-5.

40. William D. Rowley, *U.S. Forest Service Grazing and Rangelands: A History*, Texas A&M University Press, College Station, 1985, p. 12.

41. Walter Dean Burnham, "The End of American Party Politics" (Transaction) *Society*, vol. 7, No. 2, December 1969, p. 16.

42. *Congressional Record*, 51st Cong., 1st Sess., June 25, 1890, p. 6497.

43. See the article by Cuzán, "Appropriators Versus Expropriators: The Political Economy of Water in the West" in *Water Rights: Scarce Resource Allocation, Bureaucracy, and the Environment*.

44. Lamm and McCarthy, *The Angry West: A Vulnerable Land and Its Future*, p. 14.

45. Hibbard, *A History of the Public Land Policies*, p. 249.

46. *Report of the Secretary of the Interior*, 1887, pp. 9, 10.

47. *Land Office Report*, 1888, p. 41.

48. Nathaniel Pitt Langford, *The Discovery of Yellowstone Park: Journal of the Washburn Expedition to the Yellowstone and Firehole Rivers in the Year 1870*, Lincoln, University of Nebraska Press, 1972.

49. The National Park Service still classifies as "restricted" a fascinating document on this subject: Richard A. Bartlett, *The Genesis of National Park Concessions Policy*, Colloquium Paper 1980, National Park Service Archives, Harpers Ferry, West Virginia.

50. Harold K. Steen, *The U.S. Forest Service, A History*, University of Washington Press, Seattle, 1976, pp. 8-10.

51. Gressley, *Bankers and Cattlemen*, p. 71.

52. Ibid., p. 65.

53. Ibid., p. 69-70. I have altered Gressley's list slightly. On p. 71 he misnames as a Chicago investor "Duncan C. Plumb," who is actually Kansas Senator Preston B. Plumb, as becomes evident on Gressley's p. 231; I have relocated Plumb as being from the "State of Kansas," although his investments were made through a Chicago house. To the New York list I have added George Bird Grinnell. To the Boston list I have added Oliver Ames.

New York: Charles Arbuckle, John Arbuckle, Robert Bacon, James G. Bennett, Jr.,

John Bigelow, Frederick O. deBillier, William Clark, Sidney Dillon, William C. Greene, George Bird Grinnell, Rufus Hatch, Abram S. Hewitt, Augustus Kountze, Charles Oelrichs, Harry Oelrichs, Trenor W. Park, William Rockefeller, Theodore Roosevelt, Jesse Seligman, Joseph Seligman, Edmund Seymour, Horace K. Thurber, Richard Trimble, William K. Vanderbilt, Julius Wadsworth, William C. Whitney.
Boston: Alexander Agassiz, John B. Alley, Oliver Ames, Joseph Ames, Azel Ames, S. Reed Anthony, George Blanchard, Edmund Converse, James Converse, B.W. Crowinshield, George Denny, Thomas C. Durant, R. D. Evans, John H. Flagler, Augustus Gurnee, C. E. Judson, J. P. Powper, Henry A. Rice, Quincy A. Shaw, H. E. Teschemacher, Ellerton P. Whitney.
Chicago: Henry Blair, John V. Clark, John Clay, R. M. Fair, N. K. Fairbanks, John V. Farwell, Marshall Field, William Forrest, Levi Leiter, Nelson Morris, Joy Morton, Joseph Rosenbaum, Godfrey Snydacker, Portus B. Weare.
DeKalb, Illinois: Isaac Ellwood, Joseph Glidden, Jacob Haish, Henry Sanborn, John W. Gates; Troy, New York: Benjamin Mitchell, Charlemagne Tower: Washington, D. C.: Clarence King; Concord, New Hampshire: J. H. Barron; New Haven: E. G. Stoddard; Philadelphia: Isaac Seligman; Cleveland, Ohio: E. M. McGillin; Painesville, Ohio: Jack S. Casement; Akron, Ohio: David M. Goodrich; St. Louis: August Busch, Eugene Williams, R. G. Huner, C. A. Snider; Louisville, Kentucky: C. M. Tilford, H. J. Tilford, W. V. Johnson, James Moore; Oshkosh, Wisconsin: George Kellogg, Philetus Sawyer, E. C. Kellogg; State of Kansas: Senator Preston B. Plumb; Omaha: Edward Creighton, John A. McShane, J. W. Macrum; Charleston, West Virginia: Stephen B. Elkins, D. W. Smith; Indianapolis: Russell B. Harrison; Buffalo, Illinois: John B. Hunter; Peoria: Robert Ingersoll; Racine, Wisconsin: Jerome I. Case. This list is far from exhaustive.

54. Chris Emmett, *Shanghai Pierce*, University of Oklahoma Press, Norman, 1953.
55. See Ray H. Mattison, *Roosevelt's Dakota Ranches*, Bismark Tribune, Bismark, North Dakota, 1958; and Mattison, "Roosevelt and the Stockmen's Association," *North Dakota History*, XVII (April 1950), pp. 3-50.
56. Denver *Daily Tribune*, April 20, 1879 and August 20, 1879.
57. Holbrook, *The Age of the Moguls*, p. 187.
58. The literature on conservation history is enormous. The most comprehensive description of unpublished materials is Richard C. Davis, *North American Forest History: A Guide to Archives and Manuscripts in the United States and Canada*, Santa Barbara, ABC-Clio Press, 1976. The most comprehensive guide to published materials is Ronald H. Fahl, *North American Forest and Conservation History: A Bibliography*, Santa Barbara, ABC-Clio Press, 1976.
59. Joseph M. Petulla, *American Environmental History: The Exploitation and Conservation of Natural Resources*, Boyd & Fraser Publishing Company, San Francisco, 1977, pp. 169ff.
60. A good overall picture of American finance in this era is Ray Ginger, *Age of Excess: the United States from 1877 to 1914*, Macmillan, New York, 1965.
61. The story of tariffs used to fight "cheap and excessive western production" is told in Bensel, *Sectionalism and American Political Development: 1880-1980*, pp. 62ff.
62. Hutchins, *Water Rights Laws in the Nineteen Western States*.
63. Libecap, *Locking Up the Range: Federal Land Controls and Grazing*, pp. 13-30.

64. Garrett Hardin, "The Tragedy of the Commons," *Science*, December 13, 1968.

65. Brown, ed., *Reminiscences of William M. Stewart of Nevada.*

66. Leonard Arrington, *The Changing Economic Structure of the Mountain West, 1850-1950*, Utah State University Press, Logan, 1963.

67. Schurz's efforts to prevent the establishment of private property rights on the public lands may have sprung from his socialist background. Schurz was a controversial German immigrant who had fought along with Karl Marx in the Revolution of 1848, yet had come to America, won election as Senator from Missouri, first supported and then opposed the radical Republican Reconstruction program before his appointment as Secretary of the Interior by President Hayes.

In the official Soviet work prepared by the Institute of Marxism-Leninism of the Communist Party of the Soviet Union, *Karl Marx: A Biography*, Progress Publishers, Moscow, 1973, the authors on page 180 identify Schurz as the Bonn revolutionary delegate to the First Democratic District Congress of the Rhine Province in August 1848, and quote Schurz as saying that he "met face to face some of the most outstanding men of [his] day, among them Karl Marx, the socialist leader. Marx was thirty years old at the time and already the acknowledged leader of a socialist school of thought. The thick-set man with his broad forehead and dark flashing eyes, his jet-black hair and full beard immediately attracted general attention. He had the reputation of being a very considerable scholar in his own field and, in fact, what he said was weighty, logical and clear....I still remember the sharply sarcastic tone he used to pronounce the word 'bourgeois'." (Ellipsis points in the original.)

68. *Congressional Record*, House, June 26, 1888, 50 Cong., Sess. 1., 1888, pp. 5592-5593.

69. See Gene M. Gressley, "The American Cattle Trust: A Study in Protest," *Pacific Historical Review*, February 1961.

70. The board of the cattle trust included from the West, John L. Routt, governor of Colorado, R. G. Head of New Mexico, C. C. Slaughter and John T. Lytle of Texas, and Thomas Sturgis and Francis E. Warren of Wyoming; from the East, Richard T. Wilson, Jr., broker, Samuel Thomas, ironmonger, Charles T. Smillie, importer, and Charles T. Leonhardt, stockbroker, along with Nelson Morris, Chicago meatpacker, according to Gressley, *Bankers and Cattlemen*, p. 262.

71. Ibid., pp. 261-262.

72. Ibid., p. 264. The American Cattle Trust's letterbooks are preserved in the Western History Research Center, University of Wyoming, Cheyenne, and contain numerous field managers' complaints to the trustees that the Trust was overstocking allotments, disregarding water rights and range rights, usurping competing livestock lands, and displacing settlers.

73. *Buford v. Houtz*, 133 U.S. 618 (1890).

74. Ibid., at 620.

75. Ibid., at 621.

76. Ibid., at 622.

PART TWO

THE WAR FOR THE WEST

4

EVOLUTION OF THE SPLIT ESTATE

As we have seen, private rights exist in federal lands, a concept alien to many. The federal lands, often referred to as *public lands*, are in fact largely privatized. That privatization has evolved in the form of the *split estate*. In this chapter we will examine its development in detail, first going to the fundamental question

WHAT IS PROPERTY?

What constitutes property? How was it created? By what law it is validated? In what form can it be held? These questions are as old as the history of man. As long ago as the classical Greece of Plato, intellectual arguments were made for the common ownership of property. No one had a right, some asserted, to separate from the community any resource for his sole possession or occupation.[1]

Equally strong arguments were made by such classical commentators as Aristotle and Cicero that private, separate ownership of property was the only property concept compatible with a free and prosperous society.[2]

Throughout history, the question of how much control a government should exercise over property found answers ranging from the Roman concept of private ownership *in absolutum* at one extreme, to holding possession at the will of the crown as exhibited in the feudal system which dominated much of the western world up until the eighteenth century.[3]

Always at the root of such controversies lies the question; "What is property?" How can property be defined? This problem has been examined in great detail by numerous modern jurists and analysts. For example, Richard Schlatter wrote his long study *Private Property: The History of an Idea* in 1951, and it remains a standard text today. A. M. Honoré in 1961 provided a deep analysis in his essay "Ownership."[4] Dozens of scholars have added commentary. I will attempt to synthesize the most useful interpretations.

The idea that property is a *thing* one owns is deceptively simple. We understand the idea of a thing much more clearly than the idea of what it means to *own* a thing. *Ownership* is actually a highly complex concept that differs dramatically from one culture to another and from one era to another. The American system of ownership has some of its most remote roots in Roman law.

Roman property law is quite ancient, first embodied in The Laws of the XII Tables. These laws are traditionally dated at 451 B.C. when a Roman embassy reputedly went to Athens to examine the famous laws of Solon in order to settle disputes over proper legal administration back in Rome. Table VII of the Laws deals with rights concerning land, and contains details such as rules about boundaries between properties, arbitration of disputes over properties, road easements, responsibility for falling trees or rainwater damaging adjacent property, who owns fruit fallen outside a man's property and so on. So even at the beginning of Roman legal history, private property was clearly well established in both principle and practice.[5]

As Roman law developed, a vital principle emerged: the notion of ownership *absolutum et directum dominium*, absolute and direct dominion, also known as ownership *in absolutum*. Under the old Roman system a plot of land could be established as the sole possession of a private individual inviolate, even by the Roman government. This private right to the land was based upon one of two things: either *first occupation* or a *prescriptive right* created by the civil government. But by whatever means the owner originally came to possess the property, once it was legally recognized, it was his absolutely. He did with the property what he desired so long as he did not harm others with it, and even then the rules of liability were clear. The government theoretically had no power to take it from him. This was property ownership *in absolutum*.[6]

Upon the collapse of the old Republic and the rise of the Empire, rulers from Augustus onward routinely confiscated estates and bestowed them upon favorites or war veterans as a reward for service. The great Roman poet Virgil supposedly wrote his *Eclogues* in order to propitiate his confiscated estate back from Augustus. Tradition says he succeeded. So ownership *in absolutum* can only survive in the absence of a tyrant with powers of rule *in absolutum*.

Under the feudal system which dominated Europe after the fall of the Roman empire in 476 A.D., the possession of property was viewed merely as a privilege granted by the monarch or feudal lord to certain of his subjects.

We cannot properly understand the evolution of American property law and the modern storm over rangelands unless we delve a little into the particulars of feudal law.

Feudalism was based upon the military policy of the warlike Germanic tribes that poured out of the north during the final century of decline of the Roman Empire. Its essence was this: Conquering barbarian generals allotted large districts of land to the superior officers of their victorious army, who in turn dealt out smaller allotments to lesser officers and their most deserving soldiers. These allotments were called *feoda*, feuds, fiefs, or fees. The old Germanic word *fee* signified a conditional reward. The reward was the land, the condition was that the tenant should do service faithfully, both at home and in the wars, to the granting superior (the lord). The tenant took an oath of fealty (notice the relation of fealty and fee) and if he failed to live up to it by not performing the stipulated service, or by deserting the lord in battle, the lands reverted to the granting lord.

A weak king could lose his subjects' fee holding by defeat in battle while a strong king could impoverish his subjects by demanding endless funds to fight for the conquest of new lands. But feudalism proved a sturdy system. It lasted over a thousand years in western Europe and worked well as a means of securing newly conquered territories. Sharing out the spoils of war among ranking victors bound men to their leaders. Property in the feudal system has the double-edged character of a high-class bribe and an elegant threat.

Feudal property arrangements do not seem to have arrived in England until the Norman invaders placed William the Conqueror on the throne in 1066. As English feudalism developed, an elaborate law of tenures evolved, which set out those rights, duties, and liabilities of the landholder which sprang from his relationship as tenant with his feudal lord. Originally there were two chief tenures: The *freehold*, given to ranking tenants of independent means, and *villenage*, or base tenure, given to a man of servile condition but not a slave, a person annexed to the person of the lord. The types of tenures were incredibly complex. The doctrine of tenure laid down as its principal legal expression the dogma that all land is held at the pleasure of a feudatory lord, and directly or ultimately of the monarch as the supreme feudatory lord.

But it is the doctrine of estates that has seeped into the deepest cellars of American law from the feudal past. The simplest statement of the doctrine is to be found in the sixteenth-century *Walsingham's Case* (1578):

> The land itself is one thing and the estate in land is another thing: for an estate in the land is a time in the land, or land for a time: and there are diversities of estates, which are no more than diversities of time; for he who hath a fee simple in the land has a time in the land without end, or land for a time without end; and he who has land in tail has a time in land, or the land for a time, as long as he has issue of his body; and he who has an estate in land for life has no time in it longer than his own life; and so of one who has an estate in land for the life of another, or for years.[7]

Fee simple is roughly ownership, and the estate for years is the familiar lease. However, fee simple, "a freehold estate. . .absolute and unqualified, the highest and most ample estate known to the law, out of which all others are taken or 'carved,'" as *Cochran's Law Lexicon* puts it, was originally *not* ownership "absolute and unqualified."[8] We have just seen that *fee* means basically a *conditional* reward. Property cannot at the same time be both absolute and conditional, so we must look elsewhere in feudal law for absolute property ownership. That lofty status in feudal terminology was called the *alod*. The old Germanic term *alod* is etymologically related to *feod*: *od* meaning property, *fe* meaning fee or conditional reward, and *al* meaning all. Thus, *alod* meant simply "all property [rights]." Allodial property was the old Roman *absolutum et directum dominium*, or "absolute and direct dominion"—ownership in one's own right, without owing any rent or service to any superior.

The British legal authority Blackstone remarked:

> This allodial property no subject in England has; it being a received, and now undeniable, principle in the law, that all the lands in England are holden mediately or immediately of the king. The king therefore only hath *absolutum et directum dominium*; but all subjects' lands are in the nature of *feodum* or fee; whether derived to them by descent from their ancestors, or purchased for a valuable consideration; for they cannot come to any man by either of those ways, unless accompanied with those feodal clogs, which were laid upon the first feudatory when it was originally granted.[9]

Even though in Blackstone's day the king's power to interfere with private property had been largely removed by such measures as the Magna Carta, Cromwell's Revolution, and favorable land legislation, the "highest and most ample estate of ownership" of fee simple still referred back to land that only the king owned in *absolutum et directum dominium*. Today we use the term "fee simple" as if it meant "alod."

But feudal property law has bequeathed to us more than the ironic term fee simple. From it we have also inherited the notion of the split estate that bears so importantly on the storm over rangelands. The holder of feudal property was required to devote certain plots of land for crop production and other lands to serve a prescribed public service such as housing military stores or troops. The purposes for which a particular property could be legally held were multitudinous and each purpose was precisely defined in lord-tenant agreements.[10] This was the origin of the modern day *bundle of rights* concept of real estate, which is at the root of the split estate idea.[11]

American real property law came most directly from the precepts of the British jurist Blackstone. The writings of Blackstone were the foundation of English property law, reflecting his attempt to bring order to the chaotic disputes over real property which resulted from Cromwell's "Glorious Revolution" of 1688. Virtually every one of the American Founding Fathers was acquainted with Blackstone. Edmund Burke, in his famous speech to parliament on conciliation with the colonies, said, "I hear that they have sold nearly as many copies of Blackstone's Commentaries in America as in England."[12]

While the popular majorities of the day in both England and America embraced the Enlightenment view of "natural rights"—the doctrine that each individual inherently possessed certain rights, endowed by their Creator, that required no approval from society or government—strong supporters remained loyal to the system of class status which could only be created by the monarch.[13]

The American Founding Fathers were well aware of the complexity of English property law. Foremost in their minds was the determination to avoid the development of a landed aristocracy and landless vassals in the new nation. Virtually every bill of rights of the original thirteen states referred to property as a natural right of all men. The Virginia bill used the language "all men are by nature equally free and independent, and have certain inherent rights,. . . namely, the enjoyment of life and liberty, with the means of acquiring and possessing property, and pursuing and obtaining happiness and safety."[14] The doctrine was usually expressed in state constitutions as *life, liberty, and property*. Sometimes expanded, it would include references to safety and happiness. A similar declaration of the natural right of property became one of the traditional elements in American state constitutions.[15] The concept of property was so closely tied to the American Revolution that it was referred to at times as "sacred" and "the basis of all liberty."[16]

Thomas Jefferson's *Declaration of Independence* contains a clear expression of the natural rights theory in asserting that all men are created equal and endowed by their Creator with "unalienable rights" to "life, liberty, and the pursuit of happiness." But not property. Why?

PROPERTY AND NATURAL RIGHTS

Jefferson's use of "Pursuit of Happiness" as opposed to "property" in the Declaration of Independence has been a source of controversy among modern historians, but Jefferson's exclusion of the specific term "property" was in harmony with his assigned task of writing a legally correct statement justifying rebellion against King George III of England. The philosophical basis for Jefferson's choice of words is articulated in John Locke's *Second Treatise on Civil Government*.[17]

According to Locke, natural rights were expressed in two major categories. Unalienable rights were the essence of the human himself, "rights which no other man hath a right to," sometimes expressed as life, liberty, and the pursuit of happiness. Then there were alienable rights. Alienable rights involved those accoutrements of living which a free man could chose to have or not have of his own volition. These alienable rights included property rights. Property rights are alienable, i.e., a particular parcel of property can be freely transferred or alienated to another owner.

Further paralleling the theory of natural rights as set forth by Locke, the infringement of alienable rights (property) was defined as usurpation. Usurpation was deemed grounds for protest against ruling authority, but of itself, could not be a legitimate basis for the overthrow of established government. The infringement of inalienable rights, (life, liberty, and the pursuit of happiness), on the other hand, was defined as tyranny in the prevailing philosophy of natural rights, and as Locke explains, only tyranny was grounds for the overthrow of established rule.[18]

Jefferson's assigned task was to make a case against George III as tyrant. His careful choice of words in the Declaration of Independence shows his sensitivity to the political mores of his day and his understanding of the gravity of the act he was undertaking.

The American view of property as a natural right was not the invention of the Founding Fathers or of the English Blackstone. The natural right theory has been around in some form from at least the time of the Roman Republic: Cicero expounds on natural rights at some length in his *Republic* and *Laws*. The concept as articulated by the American Founding Fathers was basically a statement of justification against the feudal concept of privilege which the middle classes of the period were generally so opposed to. It was a concept developed through the centuries in the writings and thought of Hugo Grotius, John Seldon, John Lilburne, Richard Overton, Thomas Hobbes, Samuel Pufendorf, John Locke, and Adam Smith. The latter two have placed their stamp on the American concept of property more than any other economic philosophers.[19]

John Locke articulated the private right to property and clarified the natural rights theory in a manner which was both comprehensible and logical to the broad middle classes of his day. His concepts were major forces in both the American and French revolution and basis for the definition of property in the constitutions of many of our modern nations from Norway to Japan.[20]

In 1690 Locke wrote:

Though the Earth and all inferior Creatures be common to all Men, yet every Man has a *Property* in his own *Person*. This no Body has any Right to but himself. The *Labour* of his Body and *Work* of his Hands, we may say are properly his. Whatsoever, then, he removes out of the State that Nature hath provided and left it in, he hath mixed his *Labour* with it, and joined to it something that is his own, and thereby makes it his *Property*. It being by him removed from the common state Nature placed it in, it hath by his labour something annexed to it that excludes the common right of other Men.[21]

Adam Smith later expounded the labor theory of value based on the Lockean concept of natural rights. His economic classic *The Wealth of Nations*, stems from the period of the "Scottish Enlightenment." It became the "bible" of most contemporary economists. His "free market" theories are credited by friend and foe alike as having a dramatic impact on the growth and development of America and the western world.[22]

The concepts of Locke and Smith expressed themselves throughout the settlement of the American nation. The right of first use, the concept of making a portion of the "common" resource separate property is expressed nowhere better than in the practice of preemption which affected land settlement throughout our nation's history in numerous preemption acts.[23] It was federal encouragement of the western march of civilization that led to the development of first use rights to water, forage, mineral, and other utilities of the western lands. Settlers preempting unsurveyed lands mixed their labor with the land and thereby acquired a vested right in the same.[24]

WHAT IS REAL ESTATE?

Property, when we speak of land as opposed to personal property such as clothing or jewelry, is usually classified by the terms *real property* or *real estate*. This real property or real estate, in our society, still retains some of those *feodal clogs* that Blackstone referred to. Real estate in America is still viewed as an entity that can be fractionated, or split, according to its various utilities. In addition, it is still often tainted with the old feudal concept of obligations and conditional ownership.[25] When one party owns all the obligational and conditional rights to a piece of property, he possesses the entire bundle of rights, or, for the purpose of visualization, the entire bundle of sticks that makes up real estate.

Picture, if you will, a bundle of sticks tied together with a piece of twine. These sticks are each labelled with a certain property right. One stick may be labelled "surface rights," another "easement rights," others "water rights," "mineral rights," "oil and gas rights," "grazing rights," "development rights," and so on. Each right, each stick in the bundle, denotes a separate use of the property.

When the entire bundle of property rights (the entire bundle of sticks) is owned by one person or legal entity, that type of ownership is called "full fee simple" or "fee simple absolute" and the entire bundle of sticks is referred to as real estate. However, if the person who owns "full fee simple" title to the real estate decides to relinquish a certain use of that real estate to a second party—for example, if the full fee title owner decides to sell an easement for utilities or access—he has in effect sold one of the sticks from the original bundle and the result is a split estate. Two parties now have legal ownership interests in the same real estate.

THE SPLIT ESTATE

Through a quirk of history, our federal lands long ago developed numerous split ownerships, i.e., private individuals obtained a "possessory interest" in one of the sticks of the bundle of rights in the federal lands, an interest short of full fee title. Miners obtained mineral rights.[26] Ranchers obtained grazing rights. Private interests took water rights as well as easement rights, development rights, oil and gas rights and geothermal rights, etc.[27]

The federal government remained the fee title holder while many of the values in the land (sticks in the bundle) were privatized. The federal government and private individuals became joint owners of split estate lands. Two federal agencies, the Bureau of Land Management and United States Forest Service, manage a large portion of the government's residual interest in these lands.

AN ANALOGY

To understand the disorderly development of split estate ownership of the West's federal lands an analogy is helpful.

Let us use a hypothetical case which is a collage of many actual cases. In our hypothetical case let us envision an acre of swampland. Surrounding our acre of swampland are rolling hills well suited to agriculture or urban development. Urban development is far removed from our acre though, and the land is too swampy for agriculture. Over the years our acre remains largely unused and unwanted while development and growth take place in more suitable areas.

We still have to pay taxes on our acre of swampland and we are searching for any kind of use or user who can even contribute enough to help defray the cost of taxes. We would like to sell it but no one wants to buy our problems.

One day we are approached by a party who would like the right to use the water from our swampland. We allow him to use the water in exchange for help in paying the taxes. Any return from this worthless land is better than no return. Someone else is grazing a few animals out through our swamps and marshes. We charge a fee for that person's grazing privilege, a privilege we allow him to transfer to his heirs.

Later another party approaches us who would like to explore the land for minerals. We allow him the mineral rights in exchange for some improvements we want done. Someone else is willing to pay a small sum for an easement to cross our acre of swampland so he can gain access to a portion of his better land. We grant that party an easement right.

All of these people have been granted a right to the use of some utility from our acre of land. Each one of them has obtained a stick out of that bundle of sticks we call real estate. Each one of them has a legal claim to a property right interest in our acre of land. Each one can utilize our land for the express purpose for which he acquired the use right.

The man who acquired the water rights has exclusive use of the water. However, he has no right to explore for minerals or to use the newly established access road. By the same token, the man who built the access road has no right to any mineral he may discover in the process of building the road. All the while we remain the owner of record of the acre of land.

However, we are now in a radically different position than we were at the beginning. Our deed, still in our name, is now encumbered with and subject to certain exceptions. We now own a *split estate* parcel of ground.

Let's continue with our analogy.

As the years go by, urban development comes closer and closer. An industrial center appears on one side of our swampy acre, a commercial center on the other. Suddenly our worthless acre of swampland has become very valuable. Developers now see the economic feasibility of draining our swamp, hauling fill, and continuing the march of urbanization. Our once worthless land is now worth millions.

Naturally we are interested. We have always wanted to sell. Previously no one wanted more than the right to various utilities in our parcel.

The problem is, while there has been a dramatic jump in price of our once worthless parcel, every right that we released has also increased in value. The new developers, obviously, want full control of the land if they are to invest millions in it. They do not want any separate ownership of various rights in the land. They are ready and willing to pay us the high price if we will get rid of all split estate ownership. In other words, they want us to put all the sticks back into one real estate bundle.

We now face a dilemma. If we attempt to buy back the water rights, grazing rights, mineral rights, and access rights, the owner of each of those rights is in a position to hold us hostage to his demands. We either pay his price, no matter how unrealistic, to regain ownership of the split estate right, or we forfeit our chance to sell or develop the property.

To complicate matters further, suddenly we find someone claiming that he has hunting rights on our ground because he has hunted there over a period of years. We never granted that right and he has no legal document on which to make the claim, but claiming it he is. He is also taking advantage of our confused situation hoping to hold us hostage to his demands. Someone else makes a similar claim to motorized recreation rights, another to hiking rights, and someone else to fishing rights. Someone else takes note of some stray horse we have allowed to roam and graze on our acre. They now claim a right in our land as protectors of the stray horse.

The end result? The conflict created by the onslaught of legitimate and illegitimate claims to our property has prevented its sale and conversion to a more productive use. We, the owners of record, are prevented from realizing a substantial personal benefit from the property. At the same time, all those owners of split estate rights in our property can maintain the status quo. The owner of water rights continues to enjoy the use of the water. The owner of the grazing rights continues to graze. The owner of the mineral rights continues to produce minerals and the man we sold the easement to continues to enjoy free access across our acre.

While we at no time ever intended to relinquish total control of our parcel of land in fact we have lost control. While none of the parties we relinquished use-rights to have total control of the parcel, neither do we. Without intending to, we have relinquished a major portion of our interests. Each right holder has as much or more control than we do.

This analogy, when applied to the federal government, helps explain what has happened to the West's federal lands.

HOW IT HAPPENED

Our federal lands were originally owned by the United States government, the public, in full fee simple.[28] Put another way, at one time the federal government owned the entire bundle of sticks. As population pressures brought settlement and development, the federal government sold or in other ways conveyed full fee title to private individuals and entities.

The Civil War brought a new perspective to land settlement. This new perspective had little impact on the states east of the 98th meridian, but it had a very dramatic impact on the new states and territories west of the 98th meridian. When, on July 26, 1866, Congress passed an act recognizing state water appropriation systems on the federal lands, the classic split estate was created.[29] States conveyed title to water rights to private individuals as settlers put the water to beneficial use, while the federal government often retained the land the waters arose upon.

In much of the arid West the greatest portion of the real estate value was in the water. With the water rights now owned privately, this effected a legal and practical control over the land by those who owned water rights.[30]

The Act of July 26, 1866, also provided for the conveyance of rights of way for "ditches, pipelines, flumes," or other water transfer facilities across federal lands. The right of way consists of fifty feet on each side of the ditch, pipeline or flume. This right of way was private property; another stick was removed from the federal real estate bundle.*

With the conveyance of water rights, ditch, flume, and pipeline rights of way for mining, manufacturing, and agricultural purposes, other necessary developments for the related industries sprang up on federal lands. For example, stockmen who now owned water rights and ditch rights for the support of livestock grazing built living quarters, livestock handling facilities, and reservoirs. These fixtures built by private capital on the federal lands were later legitimated as private property through legislation such as Section 4 of the Taylor Grazing Act. Another stick from the federal real estate bundle had gone into separate ownership.[31]

The mining law of 1872 made possible the claiming of mineral rights. A miner could stake a claim to mineral, extract that mineral and never own the full parcel of real estate. The 1872 mining law also provided the option for conveyance of fee title to the entire bundle of sticks. Many such mining claims

* Cf. supra, pp. 68-70.

were patented (privatized) by the "location patent" method provided for in the 1872 law. Today the West is liberally sprinkled with small parcels of patented mining claims surrounded by a vast sea of federal split estate lands, but for every patented claim there are hundreds of unpatented claims where only the mineral rights are privately owned.

The most extensive of all the possessory interests created on the federal lands was the grazing right. It has also been the possessory interest most disparaged and feared by the proponents of either collectivism or monopoly capitalism. These were "first use rights" established by stockmen as they settled the West.[32] Typically the stockman located a small piece of higher quality land along the rivers and streams through a preemption act or one of the homestead acts. The government granted fee title on the preempted or homesteaded lands which became the base for a ranching operation utilizing the adjacent federal lands for grazing. He solidified his claim to grazing rights on a certain area by obtaining the rights to the water.[33] So extensive was the establishment of grazing rights in the West that quarrels arose between grazers and some sought to void the rights of others in the courts, the *Buford* v. *Houtz* case we mentioned in Chapter Three.[34] In 1890 the Supreme Court of the United States ruled that a license had been created on the federal lands and that it had been done with the full knowledge and encouragement of the federal government. This license was the permission from the federal government which many viewed as a federal endorsement of preemption as a means of recognizing private title to the range.

There is probably no item in the federal split estate that has been more controversial than the grazing right.[35] Grazing rights tied to water rights have created a significant vested interest in the federal land by those who own them.[36] It has been grazing rights more than any single stick from the federal real estate bundle that has maintained the openness of the West. As we shall see, the grazing right controversy holds the key to the future of the federal lands and their burgeoning potential contribution to the nation's economic, social and political well being.

I earlier mentioned oil and gas leases and geothermal leases as segments of the split estate. The Interior Board of Land Appeals ruled in the case of *John Bloyce Castle* on May 22, 1984 (IBLA 84-88) that an oil or gas lease on federal lands was a property right entitled to full due process of the law.

We could talk about federal leases for gravel or building material extraction and other separately owned uses on the federal lands such as recreation concessions. The federal lands of the western United States, managed by the Bureau of Land Management and the Forest Service, are not *public lands* in any realistic sense. The public has an interest in these lands only to the extent of the residual values still owned by the federal government. Private interests, on the other hand have long since claimed legal title to most of the utility values in the land. The resulting conflict between public and private interests in the same land has been the source of tremendous waste, but it also holds the key to the solution of the federal lands problem.

ATTEMPTS AT RESOLUTION

Numerous attempts have been made over the last one hundred twenty years to draw attention to the split estate-private protected property rights issue and rectify the problem of title. Major attempts to solve the federal lands problem were made during the Roosevelt administration in 1934, the Truman administration in the mid forties and the Reagan administration in the early 1980s.[37] Every attempt faltered and died.

There is a reason for the failure. These federal lands are commonly viewed, and continually referred to, as *public lands.* The term *public lands,* as commonly perceived by the public, indicates lands owned entirely by the federal government and by extension the citizens of the United States who supposedly own equal shares of the federal government's assets. Certain federal lands do meet this criterion such as some portions of our national parks and some portions of some game refuges. These are areas where the federal government owns all of the values in the land. However, even in national parks private concessioners own a possessory interest in their visitor facilities that is fully protected by the Concession Policy Act of 1965. When privatizing the public lands is suggested, many citizens envision the loss of our parks and wildlife refuges. Their reaction is predictably negative, a fact not lost on the opponents of privatization.

However, the vast majority of our federal lands do not meet the criterion of total federal ownership in fee. Most of our federal lands, particularly those under the management of the Bureau of Land Management and the Forest Service, are split estate lands. Lands in this category constitute nearly one third of the land area of the United States.[38]

A second reason every attempt to date to solve our costly federal lands problem has failed is that while some have acknowledged the existence of the split estate they have failed to recognize how numerous the split estate values are. They have suggested blanket solutions where only individually tailored solutions could apply.

For example, the Sagebrush Rebellion attempted to solve the split estate problem during the early Reagan Administration by proposing the transfer of federal lands to the states. Sagebrush Rebels acted too much from personal antagonism toward their federal masters and too much from a naive belief that changing masters could make one free. Rebels elevated the U.S. Constitution's guarantee of states' rights into an ideology that was in some way supposed to supply all the answers, although in fact it never even asked the questions. Interior Secretary James Watt had an easy job picking off Sagebrush Rebels with his "Good Neighbor" policy because by the time he took office most Rebels had figured out that the states might be an even worse master than the federal government. And no Sagebrush Rebel even thought of a practical way to transfer all the millions of split estate ownerships belonging to miners, grazers, loggers, cabin owners and all the rest into state ownership. In short, the Sagebrush Rebellion failed because it had only an ideology and no real philosophy.

The dilemma the government faces on the federal lands today is the same as the owners of our hypothetical acre of swampland. We, the owners of the swampland in our analogy, found that while technically we were still the landlord of the property, we had relinquished so many rights to the use of our land that practical and legal control had been lost.

The federal government long ago lost practical and legal control of the West's federal range lands. The transfer of water rights, mineral rights, grazing rights, easement rights, oil and gas rights, etc., served the purpose of getting a return from the land while waiting for the population pressures to eventually increase the value. Releasing certain utility rights in the West's federal lands while refusing to convey title was an effective way to placate western interests who wanted control of their resource base and *equal footing* with the East. It allowed for established sectionalist interests to maintain control of the West's economic and political development.[39] It appeared to be a way for northeastern political and economic interests to have their cake and eat it too. No one, it appears, planned for or expected the confused state of affairs that exists on the West's federal lands today.

Acknowledging the real value of the privately owned portions of the federal split estate would expose the fact that the federal government has little, if any, legal residual ownership in many of these lands. And these lands, like our hypothetical acre of swampland, have now taken on tremendous value. Resource development interests worldwide view the federal lands of the western United States as a resource storehouse of immeasurable worth. They are no longer the lands no one wants. While multinational corporations cast a covetous eye on the resource riches of the West's federal lands, they are aware there is a cloud on the title of these lands. The political and economic consequences of acknowledging that reality will be tumultuous, to say the least, but it must be dealt with before the economic stranglehold on one third of our nation's land can be broken.

The federal land management agencies have for the past two decades followed a policy of attempting to deny the existence of private rights in the federal lands. They have attempted to eliminate private rights by one device after another, the establishment of wilderness areas, the protection of wild horses and burros, and the creation of many other special designations that appeal to "the public good" but in fact act to destroy existing private rights. The agencies have attempted to encumber private rights in the federal lands with an *overkill of regulations*. This attempt to render existing private rights uneconomic has for all practical purposes become the major focus of current federal land policy.

Chapter Four Footnotes

1. Plato was an early utopian advocate of benevolent dictatorship, proposing redistribution of landed wealth, as noted at length in Karl R. Popper, *The Open Society and Its Enemies*, vol. 1, *The Spell of Plato*, Princeton University Press, Princeton, 1962. Plato's relevant works are best studied in the Loeb editions: *Republic*, translated by Paul Shorey, 2 vols.; *Laws*, translated by Rev. R. G. Bury, published in America by the Harvard University Press, Cambridge.

2. See the Loeb edition, Cicero, *De Officiis*, translated by Walter Miller, II, XXI, p. 73. Aristotle, *Politics*, translated by H. Rackham, II, p. 10. Published in America by the Harvard University Press, Cambridge.

3. For an extensive history of property see Richard Schlatter, *Private Property: The History of an Idea*, Rutgers University Press, New Brunswick, 1951. The dissenting view is held by Marxists as represented in Frederick Engels, *The Origin of the Family, Private Property and the State* (1884), International Publishers, New York, 1972.

4. A. M. Honoré, "Ownership," in *Oxford Essays in Jurisprudence*, ed. A. G. Guest, Oxford University Press, Oxford, 1961, p. 142.

5. See the Loeb edition: *Remains of Old Latin*, vol. III, *Lucilius* and *Laws of the XII Tables*, translated by E. H. Warmington, Harvard University Press, Cambridge, 1938, pp. 467ff.

6. Also called "Dominium directum et utile," see *Black's Law Dictionary*, Fifth Edition, West Publishing Company, St. Paul, 1979, p. 436.

7. A full discussion of Sir Francis Walsingham and Walsingham's Case can be found in Conyers Read, *Mr. Secretary Walsingham*, 3 volumes, 1924, reprinted Knopf, New York, 1967.

8. *Cochran's Law Lexicon*, Fourth Edition, The W. H. Anderson Company, Cincinnati, 1956, p. 136.

9. Blackstone, *Commentaries* ii, pp. 2-3.

10. *Das Urbar, Der Grafshaft Ravenberg, Von 1556*, [The Cadastral Register of the Earldom of Ravenberg, from 1556], Bearbeitet von Franz Herderhold, Band I Tex, Aschendorffsche Verlagsbuchhandlung, Munster, 1960.

11. For a discussion of the "bundle of sticks" metaphor and the general concept of split estate ownership see Alfred William Brian Simpson, *An Introduction to the History of the Land Law*, Oxford University Press, Toronto, 1961, pp. 163-94.

12. Edmund Burke, "Speech on Conciliation with America," 1775, in *The Works of Edmund Burke*, 12 vols., Little, Brown and Company, Boston, 1899.

13. For a study contrasting feudal with American concepts of property, see William B. Scott, *In Pursuit of Happiness, American Conceptions of Prosperity from the Seventeenth Century to the Twentieth Century*, University of Indiana Press, Bloomington, 1977.

14. For a discussion of the Virginia Bill's meaning see Schlatter, *Private Property: The History of an Idea*, pp. 188-89.

15. Ibid., pp. 189-193.

16. Ibid., p. 195.

17. See Chapter V, "Of Property," in John Locke, *The Second Treatise of Government*, first published 1690. I have used the edition of Thomas P. Peardon, Macmillan Publishing Company, New York, 1985.

18. See Chapter VIII, "Of the Conveyance of Adam's Sovereign Monarchical Power," in John Locke, *First Treatise of Government*, first printed in London in 1690 by Awnsham Churchill. I have used the edition of Peter Laslett, Cambridge

University Press, Cambridge, 1960.

19. For a discussion of the role of each of these legal scholars, see Schlatter, *Private Property: The History of an Idea.*

20. The world's debt to Locke for private property law is discussed in C. J. Czajkowski, *The Theory of Private Property in Locke's Political Philosophy*, University of Notre Dame Press, South Bend, 1941.

21. Locke, *The Second Treatise of Government*, Chapter V, "Of Property."

22. Adam Smith, *An Inquiry into the Nature and Causes of the Wealth of Nations*, first edition, London, 1776.

23. Hibbard, *A History of the Public Land Policies*, p. 170.

24. The actions of actual settlers preempting land is examined critically in Feller, *The Public Lands in Jacksonian Politics*, pp. 194-198.

25. Eric T. Freyfogle, "Land Use and the Study of Early American History," in *Yale Law Journal*, Vol 94., p. 717, 1985, at p. 725.

26. 17 Stat. 91-96, "Mining Act of 1872," codified at 30 U.S.C. 601-603.

27. House Committee on Public Lands, *Hearings on grazing, homesteads, and the regulation of grazing on the public lands*, 1914, p. 31. See also Interior Board of Land Appeals, Case No. 84-88, May 22, 1984.

28. Hibbard, *A History of the Public Land Policies*, pp. 7-14.

29. *U.S. Statutes At Large*, vol. 14, 39th Congress, Sess. I, Chap. 262, "An Act granting the Right of Way to Ditch and Canal Owners over the Public Lands, and for other Purposes," pp. 251-253.

30. Charles H. Shinn, "Work in a National Forest," *Forestry and Irrigation*, vol. 13, November, 1907, pp. 590-97.

31. 48 Stat. 1269, "Act of June 28, 1934," codified at 43 U. S. C. 315 *et seq.* Commonly known as the Taylor Grazing Act.

32. Rowley, *U.S. Forest Service Grazing and Rangelands: A History*, pp. 8-11.

33. Ibid. pp. 13-14, 19.

34. *Buford v. Houtz*, 133 U.S. 618 (1890).

35. Rowley, *U.S. Forest Service Grazing and Rangelands: A History*, p. 59.

36. C. R. Anderson and John A. VanWagenen, *A Report to the Committee on Appropriations U.S. House of Representatives on the Water Rights Policy of the Bureau of Land Management Relating to the Grazing Management Program*, Surveys and Investigations Staff, House Committee on Appropriations, November 15, 1984. The Report is marked "NOT FOR RELEASE UNTIL AUTHORIZED BY THE CHAIRMAN." The Chairman had not released the report as this book went to press in 1989.

37. The Taylor Grazing Act of 1934, the combination of the Grazing Service and the General Land Office into the Bureau of Land Management in 1946, and the Privatization campaign of 1981 all fell short of yielding full title to the land for the holders of prior rights. See *Privatization: Toward More Effective Government*, Report of the President's Commission on Privatization, David F. Linowes, Chairman, Washington, D.C., March 1988, pp. 242-243.

38. No precise inventory of split estate lands has been performed because the federal government does not officially recognize the split estate. However, some estimates were computed by the Public Land Law Review Commission and can be found in *One Third of the Nation's Land: A Report to the President and to the Congress by the Public Land Law Review Commission*, Washington, D. C., June 1970, pp. 105-118.

39. See Robert H. Nelson, "Why the Privatization Movement Failed," *Regulation*, July/August 1984.

5

THE FOREST RESERVES

The forest reserves that evolved into our present vast system of national forests were authorized by an 1891 law no one understood—even at the time it was enacted. Today it is generally called "The Creative Act of 1891" or "The Forest Reserve Act of 1891," although in fact it was a general revision of the land laws and repealed several acts, including the Timber Culture Act of 1873 and all preemption laws.

It was the notorious Section 24 of that Act which gave the President power to unilaterally withdraw from disposal portions of the public domain as forest reserves so that settlers could not enter.

Section 24 had been added as a last-minute rider to "An act to repeal timber culture laws, and for other purposes." This is the law that Gifford Pinchot, the first chief of the Forest Service, in his autobiography called "the most important legislation in the history of Forestry in America...[and] the beginning and basis of our whole National Forest system."[1]

The momentous rider had been tacked on in a House-Senate conference committee—but was not referred back to the originating Public Lands Committee of either the House or the Senate, a completely illegal procedure. It went straight to a floor vote and nearly every commentator says that Congress passed this most important law without being aware of its content.

Who added Section 24? What interest group of the time—northeastern capitalists, forest preservationists, professional foresters, anti-monopolists, or sectionalists—pressured for its addition, and who yielded to the pressure? Why was Section 24 added? What was the "congressional intent" behind these epoch-making forest reserves? Aesthetic preservation? Arrest of timber thieves? Preventing damage to watersheds? Assurance of future timber supplies? Thwarting speculators? Suppression of western development to guarantee eastern economic dominance? An attack on preemptive property rights? There is no convincing answer. Yet the bill passed both chambers without difficulty and became law March 3, 1891.

The language of the law presents problems of its own. Here's what it says verbatim:

> Sec. 24. That the President of the United States may, from time to time, set apart and reserve, in any State or Territory having public land bearing forests, in any part of the public lands wholly or in part covered with timber or undergrowth, whether of commercial value or not, as public reservations; and the President shall, by public proclamation, declare the establishment of such reservations and the limits thereof.[2]

That's it, the puzzling 68-word clause that generated our vast national forest system. It gave the president power only to *proclaim* forest reserves. It did not confer the power to *administer* the reserves thus created nor did it provide *appropriations* for management of any kind. Even more baffling, the first forest reserve created under Section 24 was not even related to forestry, but added a huge section to Yellowstone National Park by presidential proclamation after Congress had just rejected a bill proposing the identical enlargement! Section 24 has bewildered historians for almost a century.

Dr. Harold K. Steen, executive director of the Forest History Society, pointed out that the clause is so poorly drafted it doesn't even make sense grammatically: The sentence lacks a necessary noun, it doesn't say *what* the President may "set apart and reserve" "as public reservations."[3] However, the gist is clear: The President may reserve any public land he pleases for any purpose he pleases.

Dr. Steen wrote in his book *The U.S. Forest Service: A History*, "Much of the original documentation has been lost for what is now called the Forest Reserve Act of 1891. It is unfortunate that one of the most important legislative actions in the history of conservation is so obscure."[4]

At least half a dozen improbable people have taken credit for attaching the rider. At least half a dozen improbable explanations for its intent have been offered. The controversy over Section 24 is as confusing as the clause itself.

The idea of creating forest reserves goes back to the 1870s, so it was nothing new when Section 24 was passed in 1891.[5] Numerous newspaper stories had been written advocating reserves.[6] More than 200 forestry bills had been introduced in Congress during the twenty years from 1871 to 1891 and they all failed.[7] Most were intended to protect watersheds and attract rain. It was then generally believed that forests influenced the climate, bringing rain. Numerous rainmaker schemes were proposed to stipple the Great Plains with trees for the farmer's sake. Cutting timber was seen as an invitation to drought.[8]

But the reservation clause is thought by historians to have originated in 1889 with the law committee of the American Forestry Association (AFA), a professional organization founded in 1875. AFA's law committee consisted of three distinguished professionals: Bernhard Eduard Fernow, a German forester who had immigrated to the United States and in 1889 was the chief of the Division of Forestry in the U.S. Department of Agriculture; Nathaniel Egleston, the immediate past chief of the Division; and Edward A. Bowers, commissioner of the General Land Office in the U.S. Department of the Interior.[9]

This law committee met with President Benjamin Harrison in April 1889, presenting a petition advocating an efficient forest policy. The president was cordial but took no action. He may have been influenced by his son Russell B. Harrison, who was secretary of the Montana Stock Growers Association for over a decade. The president himself was keenly aware of western circumstances from his earlier role as chairman of the Senate Committee on Territories. While a senator, the president had advised his son that Congress was in a mean mood over a raging fencing controversy and if the ranch industry didn't watch its step, there might be a tax for the privilege of running stock on the unfenced public domain.[10]

The following year the American Forestry Association memorialized Congress to create forest reserves and to provide a commission to administer them. Congress likewise took no action.

The law committee tried again, this time with Secretary of the Interior John W. Noble. Fernow, Bowers and Egleston were joined by others, including John Wesley Powell, head of the U.S. Geological Survey, who earlier had led the first daring expedition down the Colorado through the Grand Canyon. Fernow impressed upon Secretary Noble his responsibility to protect the public domain.

As Dr. Steen wrote of the meeting, "Accounts vary as to who said what, but it is generally accepted that as a result of the meeting, Noble personally intervened with the congressional conference committee at the eleventh hour to get Section 24 added." If this story is true, the intent of the forest reserves was clearly forestry, i.e., the professional management of reserves for both protection and use. Historian Samuel Hays calls this "the gospel of efficiency" and sees it as a technocratic rather than ecological viewpoint.[11]

But one problem with this account is that the first forest reserve created under Section 24 was an addition to Yellowstone National Park. National parks have no connection to forestry or the efficient use of resources: they are managed by the dual legislative mandate of "public use and enjoyment" and the "preservation of lands and features in their natural condition."

However, John Ise, the historian who in 1920 wrote the first comprehensive history of U.S. forest policy, said that Noble added the rider. In *The United States Forest Policy* he wrote: "Noble, who had been influenced by Fernow and Bowers, and perhaps by other members of the American Forestry Association, asked the committee to insert a rider authorizing the president to establish reserves."

Ise based his assertion on a letter from Fernow, who replied to Ise's correspondence:

> My memory is, that at the time the story was current, Mr. Noble declared at midnight of March 3, in the Conference Committee, that he would not let the President sign the bill. . .unless the Reservation clause was inserted. Since these things happen behind closed doors, only someone present can tell what happened, Secretary Noble or one of the conferees. All we, that is, Bowers and myself, can claim is that we educated Noble up to the point.[12]

Noble's secretarial papers might tell the tale of what Section 24 actually intended and how it was actually inserted—a most promising lead. A search revealed that Noble's papers were donated to the Missouri Historical Society Library shortly after his death in 1912. However, they had unaccountably disappeared by 1928, when an author asked to examine them and they could not be found. This most promising lead proved to be a dead end.

Senate Historian Richard Baker offered a helpful suggestion: It is unlikely in the extreme that an executive branch officer could violate the separation of powers doctrine so blatantly as to thrust himself into a closed-door session of a congressional committee and make threats. The story was probably fabricated.

A footnote in Dr. Steen's book advises:

> see Herbert D. Kirkland, "The American Forests, 1864-1898: A Trend Toward Conservation," Ph.D. dissertation, University of Florida, 1971, pp. 171-75. Kirkland casts doubt upon Noble's specific role and whether Fernow was even aware that the amendment was under consideration.

Kirkland, a young doctoral candidate at the time, had dug up massive amounts of documentation that professional historians missed. Primarily through letters that have been in the National Archives for nearly a century, Kirkland showed that neither Forestry Division chief Fernow nor Interior Secretary Noble even *knew* that Section 24 had been added, much less had anything to do with putting it there.

Kirkland writes, "nothing at all appears on the Forest Reserve Act in the Division of Forestry papers until March 16, 1891, almost two weeks after it became law. These papers indicate that it was quite likely that Fernow was not even aware of this legislation until after it had passed." Yet, as a result of Fernow's statements and their interpretation by Ise and others, nearly everyone gives Noble credit for initiating the Forest Reserve Act.[13]

All evidence indicates that Noble found out about Section 24 on Monday, March 16, 1891, when Arnold Hague, an Interior Department employee in the U.S. Geological Survey assigned to Yellowstone, took the news to Noble in a private meeting.

Hague's letters in the National Archives show that he and Washington lawyer and Yellowstone advocate W. Hallett Phillips discovered the enactment of Section 24 late the previous week. Hague, an ardent preservationist, realized the implications of Section 24 in rectifying the recent crushing defeat of a bill asking Congress to expand Yellowstone National Park southward.

Hague took Section 24 to Noble, who asked him and Phillips to draft an appropriate proclamation establishing a Yellowstone Forest Reserve with the exact boundaries of the failed proposal to Congress. Hague delivered the draft proclamation to Noble on March 25. Noble forwarded it immediately to President Harrison, who signed it March 30, not two weeks after Noble found out about Section 24.

The very day he told Noble about Section 24, Hague discreetly spread the word to fellow preservationists. In a March 16 letter to park enthusiast John Nicholas Brown of Rhode Island, Hague wrote:

> The matter of setting aside the land adjoining the Yellowstone Park, which it was proposed to add to the Park by the bill, had already received some attention from me at the time of the receipt of your letter. I had seen

the Secretary of the Interior upon the matter and he seemed much interested and asked Mr. Phillips and myself to prepare the necessary papers which we propose doing. I will keep you advised from time to time if we accomplish anything of what form the matter takes. In the meantime I would suggest that you say nothing about it as it might defeat our object if it in any way came to the ears of those who are interested in the lands out there or feared that it would in some way interfere with their pecuniary schemes. We hope to get the land set aside without any newspaper discussion until the thing is accomplished.[14]

Hague appears to have been the first official to discover Section 24, which reduces Fernow's importance as a source of information about the rider. Many years later Fernow responded to a request for information from *Forest and Stream* magazine publisher George Bird Grinnell April 10, 1910, admitting

I do not know who drafted the exact clause, but I know it would never have been inserted, if we. . .had not educated the Secretary of the Interior to the propriety of this move.

Why did Noble ever get the credit in the first place? Again, Arnold Hague's letters reveal the answer. You have to realize first, however, that *Forest and Stream* magazine publisher George Bird Grinnell himself at one time owned a western ranch, was perfectly aware of the need for water rights and range rights, and was implacably opposed to extending them into actual title to the land. It is also important to realize that Grinnell is considered by at least one qualified historian to be the true "inventor" of the wilderness preservation idea, rather than Aldo Leopold, or Arthur Carhart, or Frederick V. Coville, who are variously claimed to hold that distinction.[15]

But, in an April 4, 1891 letter to Grinnell (who, as noted above, was years later to ask Fernow who wrote Section 24), Hague provided "such data as you may need in order to make up an editorial for your paper." After citing Section 24 verbatim, Hague guessed:

It was put in, I suppose, as a "sop" to those who believe in timber preservation.

So Hague didn't know who put it in or why. But he cautioned Grinnell:

In your editorial you had better give the Secretary of the Interior a little taffy for his seeing the necessity for this thing. You can also congratulate F. and S. [Grinnell's *Forest and Stream*] in the fact that the country has been set aside, and if you see fit to congratulate the undersigned, I would make no objections.

Arnold Hague was well connected to many eastern capitalists. In a letter to *Century Magazine* publisher Robert Underwood Johnson dated May 21, 1891,

Hague discusses Johnson's plan to create a politically powerful preservation association. Johnson had since 1873 been a distinguished editor of the popular *Century* magazine, founded by Charles Scribner. Johnson had also proven himself a potent lobbyist by helping legislate simultaneously for the international copyright law and for the creation of Yosemite National Park in 1890— "spiritual lobbying," he called it.

Johnson's proposed association was to agitate for the preservation of government lands as national parks, and now, as forest reserves. Hague sent Johnson a list of "a few men whom I think would be glad to take an interest in the matter," i.e., give money and time.

The list included Theodore Roosevelt, son of a wealthy glass merchant and at the time a member of the Civil Service Commission; Gifford Pinchot, scion of a wealthy Connecticut family, then beginning an association as a European-trained forester with the next man on the list; George Vanderbilt, heir to the Cornelius Vanderbilt fortune, at the time constructing the vast estate "Biltmore" in North Carolina where Pinchot would serve for a time as forester; Charles W. Deering, whose firm Deering Manufacturing (merged in 1902 with International Harvester Company) built and sold the Marsh Harvester, was a member of the Union Club of Chicago along with the cattle capitalists, and had an obvious reason for wanting 160 acre farming homesteads and no recognition of prior appropriation water rights or range rights; and Clarence King, eminent geologist and investor in the N. R. Davis Cattle Company in Wyoming and Nebraska and the Dakota Cattle Company.

Other inclusions on the list are the members of the National Academy of Sciences, Prof. William H. Brewer of Yale and the eminent botanist Prof. Charles S. Sargent. They were later to join with Hague, Pinchot, and Prof. Alexander Agassiz of Harvard, another investor in the N. R. Davis Cattle Company (and the Union Cattle Company of Wyoming), in the Forestry Commission of the Academy which recommended that thirteen forest reserves covering 21 million acres of federal land be withdrawn from settlement all at once on Washington's birthday in 1897. Johnson, the recipient of Hague's letter, and Gifford Pinchot in 1895 had together convinced the American Forestry Association to successfully press the Secretary of the Interior to establish the commission. Hague's network was connected to an exceptionally wide complex of capitalists, conservationists, and preservationists.[16]

However, the question arises concerning the new bureaucratic model of forest monopoly expressed in Section 24. Where did the idea come from? As most historians are aware, in the late 1880s government ownership of the means of production was widely discussed among intellectuals.

THE CONSERVATIONISTS AND THE CAPITALISTS

Hague wrote to Johnson in his May 21 letter in a comment on Vanderbilt, "From what I have heard of his plans he seems to take very considerable interest in forest preservation." Vanderbilt established the first forestry school in

America at Biltmore, and hired German forestry professor Carl A. Schenck to teach a one year practical course in forestry. Vanderbilt's family later sold his estate to the United States, and it became the present Pisgah National Forest, which is clear evidence that the Vanderbilts, for whatever reason, supported the forest reserve idea.

The bureaucratic model of government monopoly of forests, as embodied in conservation—and in the Forest Reserve Act of 1891—had a sound theoretical basis in the thought of the day. Fabian socialism, which got its name from the Fabian Society of England, was well known in America through the 1889 publication of *Fabian Essays* with contributions from George Bernard Shaw, Sidney Webb, Annie Besant and Graham Wallas.

The Fabian socialists were opposed to the revolutionary theory of Marxism, holding that social reforms and socialistic "permeation" of existing political institutions would bring about the voluntary development of socialism. Repudiating the necessity of violent class struggle, they saw socialism as a path to a rational technocratic form of government that nationalized only those resources necessary for social welfare and left private property otherwise intact. A scientific elite would run the bureaucracy. The "progressive" assumptions about government ownership underlying the conservation movement, as led by Fernow and Pinchot, are virtually identical to the principles of Fabian socialism. No well-read easterner in 1891 was likely to find the idea of government-owned forest reserves unusual.

THE PRESERVATIONISTS

The bureaucratic model of government monopoly of forests, as embodied in preservationism, like that of utilitarian conservation, also had a sound theoretical basis in the thought of the day. It was, strangely enough, based upon religious and philosophic beliefs. Historian Susan R. Schrepfer explains:

> The Sierra Club had been formed in 1892, during a period when many sought to reconcile theism and Darwinian evolution. One Club founder, Joseph LeConte, a University of California paleontologist, outlined for Americans a synthesis of direct agency and natural law. LeConte's world view was predicated upon the existence of a force external to the material world and upon man's ability to grasp the meaning of this force. He advanced a rational, scientific theology in which God has both a real, independent existence and a presence in nature. LeConte's idealism saw man as possessing two natures—the animal and the spiritual. With these dualities, LeConte perceived design in nature. Evolution was creation, "the conception of the one infinite, all-embracing design, stretching across infinite space." The very existence of man proved that evolution moves toward greater physical and spiritual development. From the late nineteenth century through the 1930s, such Sierra Club officers as naturalist John Muir, business leader Duncan McDuffie, professor of theology William Badè, and Stanford University President David Starr Jordan similarly influenced the organization with their conceptions of progressive designs in nature.[17]

The policy conclusions to be derived from these philosophic beliefs were several, historian Schrepfer says: if the human intellect is the highest product of directional evolution, our mastery of the world must be beneficial. Parks and reserves are therefore "temples within which to worship a universe of continuity and design."

Hence, "these men designed their parks to capture what they considered the highest examples of evolutionary momentum, such as the oldest of the coastal redwoods." So their religious beliefs influenced them to preserve only the best specimens of what nature had to offer. Their preservationist bureaucratic model of government monopoly of forests was selective.

It was only later, in the 1960s, that this religious vision was lost, and the Sierra Club leadership began to believe in a random evolution in which man was an accidental adaptation, an accident no better than a grasshopper or an amoeba. When that happened, when man was denied any special place in nature, Sierra Club policy changed to assert that nature had rights that man should not continually compromise.

In that climate of opinion, policy changed: all of nature should be saved, not just the best examples, because nature might contain answers to questions we would never learn to ask. Then the preservationists' bureaucratic model of government monopoly of forests became indiscriminate, seeking to monopolize the whole earth.

So there was a distinct, well-articulated philosophy behind the preservation movement of the early 1890s. It was selective, but just as fervent and self-righteous as that of contemporary environmentalists.*

THE CONFERENCE COMMITTEE

The question remains that if Noble didn't add Section 24, who did? And why?

Doctoral candidate Kirkland, who so brilliantly uncovered Hague's key role in throwing credit to Noble, could not penetrate further. "This writer," he concluded, "has been unable to determine who specifically drafted the forest reserve clause and attached it to the 'Act to Repeal Timber-Culture Laws.' It appears, however, that it came from someone within the conference committee rather than from Noble, Bowers, Fernow, Hague, Phillips, or the variety of other people sometimes given credit for publicizing the idea of reserving forest land."

The House-Senate conference committee that met to resolve differences between their two versions of the land law revision act obviously had inserted the illicit rider. The *Congressional Record* lists the conference committee members as:

Senate	House
Preston B. Plumb,	Lewis Edwin Payson,
Republican, Kansas,	Republican, Illinois
Chairman, Senate Public	Chairman, House Public
Lands Committee	Lands Committee

* This discussion is expanded upon at pp. 224-28.

Senate	House
Richard E. Pettigrew,	John Alfred Pickler,
Republican, South Dakota	Republican, South Dakota
Edward Carey Walthall,	William Steele Holman,
Democrat, Mississippi	Democrat, Indiana

Six men, none of them stellar names in the firmament of conservation. Yet one of them—at least one of them—is the father of the national forests. Personal biographies offer hints. As we already know, Senator Plumb was invested in western cattle ranches through a Chicago investment broker and may have been influenced by his eastern capitalist friends to promote the 1891 bill which, in addition to creating forest reserves, was also designed to extinguish prior appropriation rights.* However, accounts say he did not participate in the conference, but only appointed Pettigrew and Walthall as a subcommittee to study the land reform bill and take it to conference.[18] Walthall had been a military man, a general in the Confederate army, and his extensive correspondence reveals no interest in the public lands issue of any consequence. One person on the Senate side later claimed to have inserted the clause. Richard E. Pettigrew in his 1921 autobiography *Triumphant Plutocracy* wrote:

> Section 24 was placed in the bill at my suggestion to take the place of the timber culture law, which never had produced any timber. I had offered this section in the Senate Committee on Public Lands, but the Western Senators were opposed to "locking up" the country in forest reservations. In conference, while I had some difficulty, I secured an agreement which included this section of the bill.[19]

There is no evidence that the Senate Committee on Public Lands ever saw such a provision. The Senate debate over Section 24 indicates that it came as a surprise to everyone. In the same book he claims to have shaped several federal land laws single-handedly, (pages 13-16), and to have known personally every president from Andrew Johnson to Woodrow Wilson, (page 217). Pettigrew's political positions had their own quirks: his book complains bitterly against the Founding Fathers for drafting the Constitution "by property interests, for property interests alone," (page 165); he has high praise for Marxism: "The Russian Revolution is the greatest event of our times. It marks the beginning of the epoch when the working people will assume the task of directing and controlling industry," (page 392); he ends with a call for labor to "take possession of the government in all its branches, drive the lawyers out of office and repeal all laws granting privileges, and enact laws for the public ownership of all utilities of every kind that are now owned by corporations," (page 430).

Pettigrew was closely connected to South Dakota's Homestake Mining Company, whose California owner George Hearst—father of newspaper capitalist William Randolph Hearst—was a fellow Senator at the time.

*Supra, p. 65.

We can learn best from the *Congressional Record* what happened to the Section 24 rider after it was inserted during conference. The conference report containing the rider was debated on February 28, 1891 in both the House and the Senate.

In the Senate, Senator Preston B. Plumb, Chairman of the Public Lands Committee, submitted the conference report to the Senate and recommended that it pass (with the new Section 24). The Secretary of the Senate proceeded to read the conference report. Immediately, Senator Wilkinson Call of Florida interrupted the reading, saying he felt the conference version "should be printed so that we might all understand it before acting upon it." Plumb blandly asserted, "there is nothing in the report on any subject whatever that has not already undergone the scrutiny of this body, and been passed by this body."[20]

That was true of everything in the bill except Section 24. The Senate had never seen it before.

Upon completion of the reading of the report, which ended with Section 24, Senator Call pounced on the amending rider and said:

> I shall not willingly vote or consent if I know it to [contain] any proposition which prevents a single acre of the public domain from being set apart and reserved for homes for the people of the United States who shall live upon and cultivate them.[21]

Senator Plumb replied to Call: "No bill has passed this body or any other legislative body that more thoroughly consecrates the public domain to actual settlers and home-owners than does the bill in the report just read." The Senate immediately voted its approval of the bill. As we've seen, President Harrison used Section 24 of that bill to consecrate the Yellowstone Forest Reserve to keeping actual settlers and home-owners out forever.

In the House, a similar drama played itself out. Representative Payson, Chairman of the Public Lands Committee, presented the conference report to the House, where the Clerk began to read it. The chief objector here was Mark Dunnell, who in 1873 had originally introduced into the House the Timber Culture Act—which this bill repealed.

Representative Payson patiently answered Dunnell. The Clerk then read the whole bill, including Section 24. Dunnell regarded himself as a champion of forestry, but he vociferously opposed Section 24, feeling it important enough to merit its own fully detailed law. A number of Representatives asked questions, mostly about details that might affect their home constituencies. Payson and conference committee member William Steele Holman responded.[22]

In the middle of the debate Thomas Chipman McRae of Arkansas rose and said:

> I do believe, Mr. Speaker, that the power granted to the President by section 24 is an extraordinary and dangerous power to grant over the public

domain, and, if I could, I would move to amend by striking out that section. I would cordially vote to strike it out, and am sorry that it is in the bill.

MR. HOLMAN. What section?

MR. McRAE. Section 24. I do not believe, Mr. Speaker, in giving to any officer, either the head of a Department or the President, power to withdraw from settlement at will any part of the public lands that are fit for agricultural purposes and not required for military purposes. There is no limitation upon this extraordinary power if the land be covered with timber.

McRae continued for two more paragraphs, deeply worried about the power granted by Section 24. Then the *Congressional Record* notes:

MR. HOLMAN. My friend will remember that the bill in regard to the withdrawal of forest land is exactly the same as the bill passed last session, after very careful consideration.

A review of the *Congressional Record* for the 50th Congress, 1888, reveals the bill to which Holman referred, H.R. 7901, "A bill to secure to actual settlers the public lands adapted to agriculture, to protect the forests on the public domain, and for other purposes."[23]

The bill was introduced February 29, 1888, by William Steele Holman, Democrat from Indiana, who, significantly, was at that time Chairman of the Public Lands Committee.

The Democrats ruled the House in 1888 and a Democrat therefore chaired every committee, unlike the Republican-dominated 51st Congress which relegated Holman to ranking minority member of the House Public Lands Committee in 1891, and therefore to lowest-ranking man on the fateful conference committee.

Section 8 of Holman's 1888 bill contained an answer historians have been seeking for almost a hundred years.[24]

That the President of the United States may from time to time set apart and reserve, in any State or Territory having public lands bearing forests, any part of the public lands designated in this act as timber lands, or any lands wholly or in part covered with timber or undergrowth, whether of commercial value or not, as public reservations, on which the trees and undergrowth shall be protected from waste or injury, under the charge of the Secretary of the Interior; and the President shall, by public proclamation, declare the establishment of such reservations and the limits thereof, and may employ such portion of the military forces as may be necessary or practicable in protecting such or any other reservations, or any other public timber land from waste or injury; and all the provisions of this act, or of any law touching the public domain which relates to timber lands, shall be subordinate to the provisions of this section.

This word pattern shows that Section 8 is the original model for Section 24. It is not a mere coincidence. A number of forest reserve bills were introduced in both the 50th and 51st Congresses, and none except Section 8 of H. R. 7901 contained language even close to that of Section 24. Section 8 of H. R. 7901 is the only clause of any forest reserve bill ever introduced in Congress to contain even a similar sequence of words. And Section 8's grammar is correct.

Section 8 immediately answers several questions, the most obvious being why Section 24 is grammatically incorrect. Evidently someone in the 1891 conference committee cloned Section 24 word for word from the template of Section 8—but purged the clause which referred to "designated timberlands" because the 1891 Act made no such designation. A splicing phrase had to be added, which was evidently done under pressure of time, because no one seems to have noticed that the splice added the innocent-sounding word "in" at just the wrong place. Here is the 1888 prototype with the final words of the 1891 paraphrase shown in boldface type, the added splicing phrase shown in italics within brackets, and the discarded words shown in strikethrough.

> **Sec. 8. That the President of the United States may from time to time set apart and reserve, in any State or Territory having public lands bearing forests, [*in*] any part of the public lands** ~~designated in this act as timber lands, or any lands~~ **wholly or in part covered with timber or undergrowth, whether of commercial value or not, as public reservations,** ~~on which the trees and undergrowth shall be protected from waste or injury, under the charge of the Secretary of the Interior~~**; and the President shall, by public proclamation, declare the establishment of such reservations and the limits thereof[.],** ~~and may employ such portion of the military forces as may be necessary~~. . . [et cetera]

At least now we know the paradigm for Section 24. We can reasonably assume that someone on the conference committee knew it too. We see that the committee deliberately eliminated the power to administer the forest reserves, not wishing to entrust them to the Secretary of the Interior for some reason, and also deliberately eliminated the power to allocate military appropriations to protect the reserves with troops.

It is puzzling that the model for Section 24 contained administrative and appropriations provisions yet Section 24 itself did not. The last portion of Section 8 was omitted from Section 24 because it was inappropriate to the 1891 bill.

Holman's 1888 bill was accompanied by a full report from the Committee on Public Lands explaining exactly why Section 8 said what it said. Here are significant excerpts:

> The bill makes the classification of the mineral lands subordinate to the homestead provision, for the reason that extensive regions of the public lands contain more or less coal and iron, and in a great many instances more

or less gold and silver, and yet are fairly agricultural lands, and this committee believe should be so classified as a safe method of diffusing this public wealth among our people. It is also true that there are vast regions of the public lands where, at the present time, apparently, no agricultural crops can be raised without irrigation, and where, from the topography of the country, irrigation is impossible; yet the committee are convinced that it will be better for the country that these lands should remain the common property of our people rather than dispose of them in large tracts as has been proposed.

The Public Land Commission appointed by the act of the 3rd of March, 1879, in their bill reported to Congress on the 24th of February, 1880, proposed by the one hundred and fifty-third section of the bill that a homestead on the pasture lands of the United States should be authorized not exceeding to one person four sections, or 2,560 acres of land. This committee can not concur in any such measure. It is true that for the present these lands are apparently adapted only to pasturage, not capable of irrigation, may not be specially valuable or justify homestead settlements on them to the limited extent of a quarter section of land, but the committee find that from time to time homesteads are in fact established on this class of land, and that large districts of country west of the one hundredth meridian, heretofore regarded as desert lands, or at least only pasture lands, are now in a high state of cultivation without irrigation. They can not, therefore, see any excuse for granting even these lands to a single person in tracts exceeding a quarter section. Indeed, in the judgment of the committee, it would be better that the pasture lands should remain the common property of the whole people than to be monopolized in large tracts even at $1.25 an acre or at any other price.

PASTURE LANDS

(2) It is well known that a vast region of country west of the 100th meridian is not adapted to agriculture without irrigation, and it is also well known that much the larger part of that region of country does not admit of irrigation on account of the elevation of the lands and, for the present, the comparatively small supply of water for irrigation.

But the uplands and elevated plateaus and mountain ranges furnish the most valuable pasturage. Millions of cattle are now pastured on these fields. The definition of desert land given in the bill can not embrace those lands which, while valuable for pasturage, are too elevated for irrigation or not within the reach of water supplies. And yet, as experience demonstrates, these lands are being gradually taken up under the existing homestead and preemption laws, and by the fraudulent use of those laws or rather of the preemption law and the commutation clause of the homestead law such lands are now held in vast tracts. Indeed some of these lands within the grant to the Texas Pacific Railway (now vacated), generally understood to be an arid region, have been taken up under the preemption law at $2.50 per acre. It is not therefore safe to regard those lands as without value for homesteads. If not now valuable as such, they may be at least in the early future. . . .

As examination of the reports of the successive Secretaries of the Interior and Commissioners of the General Land Office for years past confirm the fact that our public lands, instead of being carefully husbanded for the high purpose of securing homes for our landless people and enlarging the number of our freeholders, are being steadily and rapidly absorbed under the present laws into great private possessions, to be held for speculation or permanent landed estates, thus defeating the spirit and purpose of the homestead law; . . .

It seems proper here to remark that in the preparation of this bill the purpose of making the public lands a source of revenue to the national Treasury has been wholly ignored as to lands that are available or can be made available *in securing homes for our people, so that only mineral lands, timber lands, town sites, and reserved lands made especially valuable by the improvements of surrounding lands and forest reserves are to be exempt from the homestead principle.* . . . [Italics in the original] The increase therefore of the number of homesteads in the disposal of the public lands is the leading object of this bill. The committee submit that the bill, in its purpose, general scope, and in its details, is essentially a homestead measure, and carefully guards against any form of monopoly of this the most valuable of the public wealth.[25]

When the 50th Congress adjourned, H.R. 7901 expired. The extensive documentation for the reasons behind Section 8, later to become Section 24, fell into obscurity, unused. Now it appears that Noble and Fernow and the American Forestry Association had little if anything to do with either Section 8 or Section 24.

Fernow drafted a comprehensive forestry bill himself in 1888, but he dealt with the Senate where Senator Eugene Hale of Maine introduced it, and its forest reserve language and provisions are significantly different from the Holman committee's: Fernow's bill authorized military protection, but also proposed a new bureau of forestry in the interior department, which was not part of the Holman committee proposal.

Numerous notables supported Fernow's bill, and the American Forestry Association memorialized Congress on its behalf, but it was tabled in committee in the Senate. It went to the House, but Holman's committee used it as a reference at most, because there is no evidence it was used at all in constructing the committee's own overall land reform bill.

Fernow's friend Bowers may have been more influential because the House of Representatives in 1888 requested a report he had written for the Interior Department the year before, "Plan for the Management and Disposition of Public Timber Lands."[26] Robert Underwood Johnson claims in his autobiography that Bowers wrote Section 24, but his account confuses the 1891 Forest Reserve Act with the 1897 forest administration law—the "Organic Act"—and thus loses credibility.[27]

The Bowers plan, like Fernow's bill, called for a new Bureau of Forestry in the Department of the Interior, which the Holman committee rejected in favor of military control. Secretaries of the Interior before Noble, Carl Schurz* for example, and General Land Office reports of land fraud, had more obvious influence in shaping the Holman bill overall, but the language of its Section 8 and the later Section 24 can only be traced to Holman.

The committee that drafted H.R. 7901 gave detailed attention to mineral and range lands in much the same terms that I have called to the reader's attention, even though the committee reached their own conclusions about the proper disposal of those lands. Their awareness of the problems of arid lands—high elevation, lack of irrigation water, recognition of pasture values—was clear. Their open rejection of the proposed 2,560 acre four-section ranch homestead provision of the 1880 Public Land Commission bill gives us an insight into where the Democrats, or at least those Democrats on Holman's House Public Lands Committee, stood on the question of land ownership—and *why* they stood there: They were Jeffersonian Democrats, still envisioning a bucolic nation of yeoman farmers happy on their small holdings. And they were sectionalists, still envisioning a West subjugated to the East.

As the committee wrote, "the bill, in its purpose, general scope, and in its details, is essentially a homestead measure." And as the committee was alert to add, the bill "carefully guards against any form of monopoly of this the most valuable of the public wealth." The notion of large land holdings for any reason whatsoever was anathema to the committee as it was to Jefferson. The Revolutionary War-era suspicion of great landed estates had fossilized into dogma. The reality behind their references to successful small non-irrigated homesteads was far less glorious than they made it out to be. Politics blinded them to the fact that even these few arid-land quarter-section farmers were routinely plundering nearby public timber just like the railroad and timber barons because there was no other way to make a living.

The House Public Lands Committee acted upon H.R. 7901 in 1888 because of many influences. Each was important and no single one can be accurately characterized as *the* major influence: committee members received homesteader complaints about too-small plots in arid lands, cattlemens' complaints about homesteaders, pressure from northeastern financial interests to keep railroads, timber, and mining available to their control, moralists' complaints about dummy entrymen and the evils of monopolies, letters from preservationists, memorials from the American Forestry Association and pressure from other personalities on the committee.

To further illustrate the actual legislative purpose of the forest reserves, I quote a few more paragraphs from the 1888 committee's report:

> One of the most radical changes proposed in the existing laws is in regard to timber lands. The bill provides "that lands chiefly valuable for timber of commercial value, as sawed or hewed timber, shall be classified as timber land;" that the land shall not be sold, but the timber on the land in legal subdivisions of 40 acres shall be appraised and sold on sealed

*Supra, p. 68, 78n.

proposals to the highest bidder, and shall be removed from the land within five years from the date of sale, the timber so purchased to be deemed personal property for all purposes, including taxation. But no person or corporation shall purchase or hold more than the timber on one section of land. If such lands, however, are found to contain valuable deposits of minerals they shall be subject to entry under the mineral laws, subject of course to the rights acquired by the purchaser of the timber. . . .

Your committee are satisfied by their investigations that high public interest will be subserved by only allowing the sale of timber on lands when the timber is of commercial value. This bill does not specially provide that the land after the timber is removed, if valuable for agriculture, shall be subject to homestead entry, for the reason that it is not clear what should be the policy of the Government in that respect, while the land may be subject to be embraced in a public reservation of timber land. If not so embraced, in the opinion of some members of the committee it should be subject to homestead entry. . . .[28]

Even the committee that originated the Section 8 that was transmogrified into Section 24 was not sure what should be done with public timberlands after timber harvest. And they were equally concerned with timber, mining, grazing and farming. Even their rigid anti-monopoly ideology contemplated many valid alternatives, such as sale to homesteaders, use for grazing land, and retention as future timber supply reservoirs.

SURMISES

What can we legitimately conclude from this new evidence about the origin of the forest reserves? How can we fulfill the duty of providing interpretation without stretching our still-meager facts too far?

At a minimum, we may observe that Section 8 of H. R. 7901, portions of which which became law in Section 24 of the Forest Reserve Act of 1891, had no original *primary* purpose. We can see a number of mixed purposes: timber harvest, mining, watershed protection, farming, grazing, and so on. It is not unreasonable to compare this mixed purpose with the principle we now call "multiple use."

The 1888 report bears the signature of William Steele Holman, Chairman, Public Lands Committee, House of Representatives, United States Congress. So does the bill, H.R. 7901 itself. Although we cannot say for certain that Holman *wrote* the language of Section 8—and by extension of Section 24—we can certify that he is *legally responsible* for that language as committee chairman and signator of the 1888 House report, and that he served in the 1891 conference committee that inserted Section 24.

It appears nobody in the conference cared much about Section 24. Senator Plumb had the amended bill read to the Senate and said it contained nothing they hadn't passed before—he may have confused the *House* passage of Section 8 in 1888 with the new 1891 Section 24. He was suffering terrible headaches and had been diagnosed with aphasia. He died of a stroke before the year was out.

Payson had the amended bill read to the House and Holman said it contained nothing they hadn't passed before, which was true.

Both houses passed the bill, President Harrison signed it into law, Hague read about it in a newspaper and told Noble, and Yellowstone National Park was expanded. *Forest and Stream* magazine spread the Noble "taffy tale." Fernow repeated it to historian John Ise, who repeated it to other historians. The mysterious forest reserve clause perplexed everyone so greatly that administrators and legislators made up the rules as they went. Congress finally passed an administration and appropriation measure for the forest reserves in 1897. The forest reserves went to the newly created Forest Service in 1905, and their name was changed to national forests in 1907.

OUTFALL

A one-sentence rider attached to the Act to repeal the Timber Culture Act of 1873 was used by Presidents Harrison and Cleveland to withdraw millions of acres of the West's mountain ranges from public settlement for "conservation" purposes. The real impact was counter-historic. As put by historian William D. Rowley, "by assuming the role of perpetual owner of these lands and their resources, the federal government began a reversal of a three century-long policy of land privatization on the frontier."[29]

Immediately after the establishment of the first forest reserves in 1891, because Section 24 provided nothing for administration or appropriations to manage the new reserves—and because men like Arnold Hague ran amok for the national park cause—the resources in those reserves were totally locked up. Timber cutting and livestock grazing were prohibited. Hunting and fishing on the reserves was prohibited. One could not even legally set foot within the reserves. The park preservation movement of Hague and Grinnell and the forest conservation movement of Fernow and Pinchot fit the needs of the eastern politicians and industrialists in their efforts to inhibit the growth and development of a resource-rich West.

Section 24, even though it originated as a homestead measure to encourage western development, constituted a serious attack on preemptive property rights when coupled with the repeal of all preemption laws in that same 1891 Act. Section 24 would in 1905 give rise to a politically and economically powerful bureaucracy: the Forest Service in the U.S. Department of Agriculture. The Forest Service would in many ways become the in-house federal agency of the northeastern money trusts in their fight against the ranching community of the West. The Forest Service would act to inhibit economic development of the West and to insure continued Eastern influence over the West's resource base.

Chapter Five Footnotes

1. Gifford Pinchot, *Breaking New Ground*, Harcourt, Brace, and Company, New York, 1947, p. 85.
2. 26 Stats. 1095 (March 3, 1891).
3. Telephone interview with Dr. Steen, September, 1988.
4. Harold K. Steen, *The U. S. Forest Service: A History*, University of Washington Press, Seattle, 1976, p. 26.
5. The first forest reserve bill was introduced in 1876 by Representative Greenbury L. Fort of Illinois and died in committee. U. S. Congress, House, "For the Preservation of Forests of the Public Domain Adjacent to Navigable Rivers," 44th Cong., 1st Sess., 1876, H. R. 2075.
6. A specific newspaper article advocating forest reserves can be found in the *New York Times*, September 18, 1885, p. 4, col. 4.
7. Bernhard Eduard Fernow in 1897 produced a chart for the American Forestry Association listing 93 pieces of legislation introduced in Congress from 1871 through 1890. He included 69 more from 1891 to 1897 and 52 more in a separate list of bills introduced from 1871 to 1897. See *Proceedings of the American Forestry Association*, XII, pp. 41-58.
8. The idea finally died in the late 1880s. See *New York Times*, August 5, 1888, p. 4, col. 4, "the old idea that trees directly induce a larger rainfall is now pretty much given up."
9. Andrew Denny Rodgers III, *Bernhard Eduard Fernow: A Story of North American Forestry*, Princeton University Press, Princeton, 1951, pp. 142-43.
10. Russell B. Harrison to Benjamin Harrison, October 24, 1885. Wyoming Stock Growers Association Collection, Western History Research Center, University of Wyoming.
11. Samuel P. Hays, *Conservation and the Gospel of Efficiency: The Progressive Conservation Movement 1890-1920*, Harvard University Press, Cambridge, 1959.
12. John Ise, *The United States Forest Policy*, Yale University Press, New Haven, 1920, p. 115.
13. Herbert D. Kirkland, *The American Forests, 1864-1898: A Trend Toward Conservation*, unpublished Ph.D. dissertation, University of Florida, 1971, p. 175.
14. All of Hague's letters discussed here are in Record Group 57, Hague Papers, letter book 3B, National Archives. Hague to E.M. Dawson, March 16, Hague to Brown, March 16, 1891, Hague to Noble, March 25, 1891, Hague to Grinnell, April 4, 1891, Hague to Grinnell, April 6, 1891.
15. John F. Reiger, *American Sportsmen and the Origins of Conservation*, Winchester Press, New York, 1975, p. 262, n. 23. Reprinted by University of Oklahoma Press, Norman, 1986.
16. The list in Hague's letter reads: Theodore Roosevelt, Civil Service Commission, Washington; Archibald Rogers, Hyde Park, Dutchess County, New York; W. Hallet Phillips, 1707 H Street, N. W., Washington, D. C.; Moses Harris, 1st Cavalry, Milwaukee, Wis.; Clarence King, Century Club, New York, N. Y.; Charles S. Sargent, Brookline, Mass.; Prof. W. H. Brewer, Yale College, New Haven, Conn.; C. W. Deering, Cor. Fullerton and Clybourn Avenues, Chicago, Illinois; Gifford Pinchot, 2 Grammercy Park, New York, N. Y.; Gen. B. H. Bristow, New York City, N. Y.; Hon. W. C. P. Breckenridge, House of Representatives, Washington; M. G. Seckendorff, Office of New York Tribune, Washington; George Vanderbilt, Fifth Avenue, New York, N. Y.; Edward Burnett, Southboro, Mass.; Prof. E. W. Hilgard, University of California, Berkeley, Cal.

17. Susan R. Schrepfer, "Conflict in Preservation: The Sierra Club, Save-the-Redwoods League, and Redwood National Park," *Journal of Forest History*, vol. 24, No. 2, April 1980, pp. 60-77.
18. *New York Times*, December 21, 1891, p. 1, col 6.
19. Richard F. Pettigrew, *Triumphant Plutocracy: The Story of American Public Life from 1870 to 1920*, The Academy Press, New York, 1922, p. 12.
20. *Congressional Record*, Senate, 51st Cong., 2d sess., February 28, 1891, p. 3547.
21. Ibid.
22. *Congressional Record*, House, 51st Cong., 2d sess., March 2, 1891, pp. 3611-16.
23. U. S. Congress, House, *House Resolution 7901*, 50th Cong., 1st Sess., 1888, introduced by William S. Holman February 29, 1888.
24. Holman's biography is in Israel George Blake, *The Holmans of Veraestau*, The Mississippi Valley Press, Oxford, Ohio, 1943. For a discussion of Holman's experience in the West, see K. A. Soderberg and Jackie Durette, *People of the Tongass: Alaska Forestry Under Attack*, Free Enterprise Press, Bellevue, 1988, pp. 124-128.
25. U. S. Congress, House, 50th Cong., 1st Sess., 1888, *Report No. 778*, signed February 25, 1888 by William S. Holman.
26. U. S. Congress, House, *Public Timber Lands*, 50th Cong., 1st Sess., 1888, H. Ex. Doc. 242.
27. Robert Underwood Johnson, *Remembered Yesterdays*, Little, Brown and Company, Boston, 1923, p. 294.
28. U. S. Congress, House, 50th Cong., 1st Sess., 1888, *Report No. 778*.
29. Rowley, *U.S. Forest Service Grazing and Rangelands: A History*, p. 4.

6

HOUR OF CRISIS

The authorization in 1891 of the forest reserves followed closely on the heels of the opinion rendered in *Buford* v. *Houtz* (1890). The reader will recall that the U. S. Supreme Court in that opinion held grazing to be a valid use of the unclaimed lands of the United States.*

> We are of the opinion that there is an implied license, growing out of the custom of nearly a hundred years, that the public lands of the United States, especially those in which the native grasses are adapted to the growth and fattening of domestic animals, shall be free to the people who seek to use them where they are left open and unenclosed, and no act of government forbids this use.... The government of the United States, in all its branches, has known of this use, has never forbidden it, nor taken any steps to arrest it. No doubt it may be safely stated that this has been done with the consent of all branches of government, and, as we shall attempt to show, with its direct encouragement.... Everybody used the open unenclosed country, which produced nutritious grasses, as a public common on which their horses, cattle, hogs and sheep could run and graze.[1]

This ruling unravelled the hopes of northern core capitalists for an easy way to usurp prior appropriation rights of earlier settlers. Under this ruling they could no longer argue that private water rights created on the federal lands by prior appropriation were invalid. The U.S. Supreme Court had taken away the main pillar of the capitalists' case against earlier settlers. If grazing was a valid use of the unclaimed federal lands, then it followed that grazing could be used as a means to prove beneficial use of those non-navigable waters on the federal lands which supported livestock.

In other words, the states were now able to create "property" from the federal estate within their borders if it was being used for grazing. If grazing was a valid use, how could anyone argue that preemption had not occurred with or without congressional approval? The Lockean concept of first use and appropriation, which had been basic to the settlement of the federal lands to 1890, was nowhere better illustrated than in the vast arid range lands of the West.

With *Buford* v. *Houtz* in place, eastern capitalists could no longer look to the courts for help, but had to look elsewhere to fulfill their dreams of domination. The Forest Reserve Act—as passed with no provisions for funding or administration—was an effective instrument to block prior appropriation and privatization in a single blow.

Further, since the opinion in *Buford* v. *Houtz* was handed down simultaneously with the failure of the American Cattle Trust, it was clear to the eastern capitalists that neither monopoly nor the courts could assure their objective of dominating the West. Only political manipulation could now promise results.

*Supra, p. 72-74.   115

Thus it is reasonable to ask whether the northeastern core capitalists had a hand in that Section 24 rider.

More than two hundred bills to designate forest reserves had been offered to Congress over the twenty year period prior to 1891. They had all failed. The embryonic conservation movement had clamored for forest reserves since the 1860s but the idea had no appeal in *laissez faire* expansionist America, a nation in pursuit of its manifest destiny. Beginning in the late 1880s, northern core capitalists called upon individual conservationists to lend their public endorsement to the slowing or stopping of development in the West. Robert Underwood Johnson's recruitment of John Muir to fight for a large Yosemite reservation in 1889 is an early example. Neither of them, however, appears to have been aware of the addition of the Section 24 rider until after the fact. We do have Arnold Hague's letter to Johnson—only three weeks after the forest reserve measure was passed—recommending that he contact the "list of capitalists" as discussed in Chapter Five.

The haste with which the Section 24 forest reserve rider appears to have been added to the 1891 act precluded any real discussion of its potential impact or even any serious discussion of its intent. We know from the 1888 Public Lands Committee Report that William Steele Holman viewed it as a measure temporarily holding the vast federal forest lands for eventual conveyance in small parcels to the small settler. We also know from Robert Underwood Johnson's autobiography that the publisher had successfully lobbied Holman to vote for the 1890 Yosemite National Park bill:

> One of the members of the House Committee was Judge Holman of Indiana, whose circuit had adjoined that of my father when he was Judge of Court. The two, though political opponents,—Holman being a Democrat and my father a radical Republican,—had nevertheless been friends, and I felt sure that Holman would be predisposed to the scheme. I also knew Mr. Plumb of Kansas, who was chairman of the Senate Committee, and I believed he would help. I told Muir that I thought there would be no serious objection to such a measure and that in my judgment it would go through.[2]

Consider this interesting linkage that bears on the 1891 forest reserve rider: Johnson to Muir to Hague to Hague's "list of capitalists" to Holman to Plumb. Plumb was listed by historian Gressley as an investor in western cattle ranches through a large Chicago cattle and land brokerage. That extends the linkage to Gressley's list of 93 cattle capitalists.*

The fact that the conference committee got away with adding this section without review by the respective full committees (an illegal act in itself) may be evidence that other members of Congress were also part of this linkage. The fact that the section was added with no consideration for appropriations and no administrative provisions is perhaps evidence that Section 24 was added in haste if not desperation.

Once Section 24 was passed into law, the nature preservation movement, in particular Robert Underwood Johnson's *Century* magazine, toned down the

*Supra, note 53, p. 76-77.

earlier crusade lamenting the destruction of private forest lands east of the Mississippi.* Preservationists were less critical of the northern core interests who had profited from that destruction. They now found themselves supported publicly by some of the same northern core interests—such as railroad magnate Edward H. Harriman—as long as they focused their obstructionist efforts on the resource-rich lands of the emerging western states.

It soon became apparent that grazing was to be the main target of the forest reserves. Grazing was the single use that could block eastern dominance through the prior appropriation doctrine as upheld in *Buford* v. *Houtz*. In February of 1893 *Century* magazine published an article on the reserves by John Muir. Muir used his growing celebrity to attack livestock grazing. While deploring cattle in general and sheep as "hoofed locusts," he declared, "One soldier in the woods, armed with authority and a gun, would be more effective in forest preservation than millions of forbidding notices."[3]

The next year, on April 14, 1894, the Department of Interior issued its first official statement of administrative policy on the Forest Reserves: a total curtailment of grazing. It prohibited the "driving, feeding, grazing, pasturing, or herding of cattle, sheep, or other livestock," on the forest reserves.[4] Whatever property rights the states thought they had were being denied to them on the richest of their resource lands. Significantly, the government issued no policy statement on timber protection, the ostensive purpose of the "forest" reserves.

In 1895 Robert Underwood Johnson sponsored a *Century* Symposium in which leading conservationists Charles Sargent and John Muir recommended the use of troops to prevent "all types of trespass," particularly grazing.[5] The nature preservationists directed their rhetoric toward the one class of users that held the potential to repeal the reserves. Ranchers with their claim to ownership of the water on these reserves by virtue of proof of beneficial use through livestock grazing were the prime target. Conservationists claimed that livestock grazing threatened tree growth, increased threat of fires, and degraded watershed for irrigation and urban purposes.

This was simply a smokescreen to avoid facing the split estate. The capitalist sponsors of the forest reserve movement wanted to clear title on these most resource-rich federal lands. They wanted to void the states' ownership of the water—held since 1866—and they wanted to void protected private property rights on the land as based upon recognition of livestock grazing as proof of beneficial use of the water. As more and more reserves were created by President Harrison and then President Cleveland, the need for a politically palatable explanation for the withdrawal of millions of acres of resource rich lands from settlement and development grew. There had to be an acceptable reason to plunder the western states' resource base. Again, the capitalists used their influence on the nature preservation movement to gain respectability for their objectives.

At an 1895 American Forestry Association meeting, Gifford Pinchot—who had been trained as a forester at L'Ecole Nationale Forestière at Nancy, France—joined with Robert Underwood Johnson and talked the membership into

*Cf. supra, p. 67.

forming a special forestry commission to be made up of members of the prestigious National Academy of Sciences. Bernhard Fernow protested that action, not a commission, was now necessary, but the Association ignored him. The commission was to put the esteemed stamp of scientific authority on the forest reserves. The Association pressured Secretary of the Interior Hoke Smith into requesting the Academy to convene such a commission. The National Academy of Science Forestry Commission was established on February 15, 1896, to be chaired by Harvard Professor Charles S. Sargent.

Professor Sargent advocated the training of forest officers at the U.S. Military Academy at West Point. He also proposed a Forest Experiment Station near West Point and the enlistment of a forest guard unit to carry out the principles of applied forestry.[6] Such proposals provided a timely and politically palatable cover for the government's blatant attack on the rancher's protected private property rights—rights which had been created through state authority according to local law and custom and rules of the court.

Members of the Forestry Commission included Prof. William H. Brewer of Yale, geologist Arnold Hague, Prof. Alexander Agassiz of Harvard—a heavy investor in the N. R. Davis Cattle Company and the Union Cattle Company of Wyoming—Army engineer Henry L. Abbott, and Gifford Pinchot, who had just left the employ of George W. Vanderbilt as a consulting forester and was the only member who did not belong to the Academy. The Forestry Commission traveled West, made its study, and recommended that thirteen forest reserves covering 21 million acres of federal land be withdrawn from settlement all at once on Washington's birthday in 1897.[7] These men were not only well-connected to the northeastern core cattle capitalists, one of them *was* a northeastern core cattle capitalist.*

THE WESTERN STATES REACT

The western states greeted the forest reserves with fury. Once the full realization of the plunder being worked upon their resources sunk in, many western politicians sought the immediate repeal of the reserve act. After President Grover Cleveland added over 21 million acres to the forest reserves by proclamation on February 22, 1897 most westerners denounced the entire forest reserve system.[8]

Cleveland had lost his race for renomination to the office of the presidency. He blamed his defeat on the controversy over bi-metalism, an issue dear to the hearts of many westerners in 1896.[9] Many felt Cleveland's forest reserves were created as a last minute act of vengeance against the western silver interests prior to leaving office.

Bitter, seething resentment appeared in statements of western congressmen. Congressman John Wilson of Washington asked if the West was to be everlastingly and eternally harassed and annoyed and bedeviled by "these scientific gentlemen from Harvard College" (referring, of course, to the National Academy of Sciences Forestry Commission). Congressman Charles Hartman of Montana called President Cleveland's action "a parting shot of the worst enemy that the American people have ever had." Congressman Freeman

*Cf. supra, p. 64, p. 100.

Knowles of South Dakota referred to the action as a "villainous order" saying, "We know the 'rotten buroughs of the West,' as the *New York World* calls us, have little influence with this administration."[10]

William S. Rowley in his book *U.S. Forest Service Grazing & Rangelands* relates the responses of westerners to what they perceived as a clearly intolerable situation:

> During the last days of the Cleveland administration, western congressmen determined to bring the new reserves back into the public domain. An amendment to the Sundry Civil Appropriations Bill, giving the president authority to modify or rescind any previous executive order creating forest reserves, passed. The entire appropriations bill including this amendment was sent to President Cleveland during the last days of his administration. The president killed the entire appropriations bill with a pocket veto, leaving the government temporarily without operating funds. The protestors would have to wait for the new administration to obtain relief from the national forest-reservation proclamations.
>
> To obtain an appropriations bill for government operations, President McKinley called a special session of Congress shortly after his inauguration on March 4, 1897. An amendment to this new appropriations bill by Congress achieved a compromise between anti-reserve forces and those both inside and outside the West who supported an active governmental forest reservation policy.[11]

This amendment made some concessions to anti-reserve western interests. Provisions in the amendment gave settlers, miners or residents free timber and stone for use as fire wood, fencing, and building material. The amendment provided that the reserves would be open to mining and prospecting.

However, no concessions were made to the stockman. As Rowley explains, "The previous condemnations of grazing in the forests had been so pointed and strident that to mention this use at all would have appeared as a total concession to it."[12] Stated another way, any concession to livestock interests would have been viewed as an open acknowledgement of the ranchers' preemptive range rights.

It is revealing to examine the complaints written by private citizens in the West. A series of statements written in 1902 protesting the Blue Mountains Forest Reserve in Grant County, Oregon, demonstrates the outrage of local citizens.

Ranchers B. C. Herburger and W. S. Southworth said about north central Oregon, "It contains no timber of any value and it would be safe to say that there is not to exceed $100 worth of timber in the whole township. The settlers in this township are all owners of stock. . . . To be required to secure a permit in the manner prescribed by the regulations of the General Land Office would compell [sic] a large number of these stock men to retire from the business."[13]

Grazing rights rather than timber were often the interest most affected by

Forest Reserves. Six residents of Grant County, Oregon made the point: "The entire township is what is called table land and is devoid of timber except in the extreme heads of the canyons where some scrubby 'Bull' pine is found, but in no instance is it of commercial value. . .but the entire township is valuable grazing land."[14]

The commissioners of Grant County protested the forest reserve's adverse economic impacts:

In the Matter of the	)
Establishment of a Forest	)
Reserve within Grant County	) Protest against Forest
Oregon, by the General	) Reserve
Government of the	)
United States	)

Whereas, By an order of the Hon. Secretary of the Interior, made and entered on the 28th day of July A.D.1902, more than one half of the total area of Grant County, Oregon, was withdrawn from settlement and occupation and set aside as a forest reserve, and

Whereas, By the withdrawal of said lands from settlement and occupation the timber supply of our County will not be sufficient to supply the demands for lumber and building material and for fuel necessary in the growth of our Towns and for the development of our agricultural and mining industries, and

Whereas, by making this forest reserve permanent upon the lines established by the said order of the Hon. Secretary of the Interior, every business industry of our County will be impaired and extra burdens imposed upon the tax payers of our County,

Therefore we, R. R. Mc Haley, County judge and P.P. Kilbourne and W.S. Brown, County commissioners, in session assembled as the County Court of Grant County, Oregon, do most earnestly protest against the establishment of a permanent forest reserve upon the lines designated by the Hon. Secretary of the Interior in his order of July 28th 1902. and demand from our General Government a just recognition of the business interests of our County.

R. R. Mc Halley, County Judge.
P. P. Kilbourne)
W. S. Brown   ) County Commissioners[15]

Another sworn statement signed by four residents of Grant County concluded, "the retention of this township in a permanent reserve would greatly retard the mineral wealth, and seriously inconvenience all resident farmers and stockmen and deprive a number of settlers of their homes. . . ."[16]

In a letter dated October 30, 1902, a miner who had traveled to Portland with a mining delegation to Senator John H. Mitchell stressed the adverse impact that the mining economy would suffer under the guise of "protecting timber in forest reserves:"

[T]his delegation opposed the Blue Mountain Forest Reserve in toto, that is to say, that it was opposed to the whole reserve and united in the request that no final action should be taken on any part of it until such time as the people residing in the proposed reserve could be heard, and their rights and interests be made known to the department. . . .

The actual and bonafide residents within that portion of the reserve lying in the counties of Grant, Baker and Harney were never consulted about the proposed reserve and never heard of it, and knew nothing about any contemplated actions, until long after the order was made . . . and nobody yet claims to know how, when, or upon what information the order of withdrawal was made, or why it was made in the manner in which it was. . . .

Within the borders of the proposed reserve, in Baker County, 2633 locations of quartz mining claims and 439 locations of placer claims have been made since 1886. In addition thereto, 45 different quartz mining patents have been issued. . .making a total of 3,250 mineral locations, embracing an area of 65,040 acres of mineral land in Baker County, within the limits of the proposed reserve. . . . From such facts alone, it would clearly appear that a large portion of the proposed forest reserve is mineral land. . . .The United States law, Act of June 4, 1897, expressly provides ". . .but it is not the purpose or intent of these provisions of the Act Providing for Such Reservations to authorize the inclusion therein, of lands more valuable for the mineral contained therein and for agricultural purposes, than for Forest Reserve purposes". . . . It is very apparent that prior to the issuing of the order, no investigation was made; or, in the event an investigation was made, that the facts were either concealed or misrepresented. . . .

[The miners in the delegation] knew that in effect the whole mineral belt of Eastern Oregon was included and embraced within the proposed forest reserve; that it was done without their knowledge, or the knowledge of any persons residing within the mineral belt. They knew from the statements and representations of leading and prominent citizens of Grant and Harney Counties that such counties were opposed to the forest reserve. . . .

[T]he whole spirit of the law creating forest reserves was violated by including therein the mineral belt of eastern Oregon. . . . Mr. Sudworth [George B. Sudworth of the Forest Bureau of Washington D.C.] frankly conceded that the law creating forest reserves did not apply to land which is more valuable for the mineral therein than for forest purposes.[17]

The impact of the forest reserves on the economic development of the West was noted by the Washington state legislature when it passed House Memorial No. 6 in 1907. They viewed the forest reserves as retarding the settlement of the West. They deplored the confiscation of property rights which had previously been acquired in forest reserves. They criticized the forest reserves for being used to attack state control of water rights and to impede and discourage livestock grazing. The entire memorial is worth reviewing because it so clearly airs contemporary attitudes on Section 24.

HOUSE MEMORIAL NO. 6.

To his Excellency, Theodore Roosevelt, President of the United States of America; To the Honorable Senate and House of Representatives of the United States; and to the Honorable Secretary of Agriculture:

We, your memorialists, the Senate and House of Representatives of the state of Washington, in legislative session assembled (tenth regular session), most respectfully represent and pray as follows:

Whereas, the Congress of the United States passed an act entitled "An Act to provide for the entry of agricultural land within forest reserves", approved June 11, 1906 (34 Stat. 233); and

Whereas, said act, while beneficial to a limited extent, the same is wholly inadequate to accomplish the purposes for which it was enacted. Under said act, a prospective homesteader (who is generally a man of limited means) coming from the East to the West to make settlement upon lands within the forest reserve, is not permitted to select for himself non-mineral land, such as to him is satisfactory for a permanent home, but is subjected to long delays, lasting months or years, before he can make his settlement, during which time there is an element of uncertainty as to whether or not he will ever be able to select a homestead desired by him; and as to whether or not a given piece of land is suitable for a homestead must be determined by some subordinate of the Agricultural Department, who, perhaps, is wholly ignorant of local needs and necessities; and

Whereas, boards of trade, commercial bodies, "booster" clubs and railway companies have distributed millions of pages of literature to induce settlers to acquire homes of their own on the public lands of the United States under the homestead laws: and

Whereas, in response to said advertising, tens of thousands of families have deserted their homes in the East, and now find themselves in a new country among strangers, with limited means, and are confronted with the fact that hundreds of thousands of acres of public land suitable to be taken for homesteads have recently been withdrawn from settlement as forest reserves; and

Whereas, now the serious question arises: "what shall such citizens and their families do?" and

Whereas, it is a gross injustice to make wholesale withdrawals of public lands from homestead settlement under the guise of "temporary withdrawal" with no regard as to whether or not such lands embrace timber or prairie lands, or lands necessary for forest reserve purposes; and

Whereas, large portions of such lands so withdrawn are essentially agricultural lands; and

Whereas, the making of forest reserves out of lands less than 4500 feet in altitude above sea level retards and prevents the settlement and prosperity of the west, by reason of the fact, among other things that home builders in the forest reserves will be isolated, without hope of near school or church privileges; without reasonable expectation of the making of suitable roads or the keeping of them in proper repair without extortionate individual expense and running the gauntlet of "red tape"; and

Whereas, for like reasons, to a large extent, there is a confiscation of the property of those who have heretofore acquired rights or titles to lands now within the forest reserves, it being a matter of common knowledge that the denser the population in a given community, the higher is the price of land; and conversely, the more isolated, the less valuable; and anything retarding or preventing settlement correspondingly decreases the value of the land; and

Whereas, there are large numbers of settlers who, for years, have resided upon unsurveyed lands who now find themselves within boundaries of forest reserves, who if from sickness, or other unavoidable casualty are compelled to leave their homestead claims, will virtually lose the result of years of toil and their improvements, for the reason that when they quit their premises before making final proof, their lands will revert to the forest reserve; and in most instances no purchaser of the improvements can be found who is willing to become entangled in the meshes of forest reserve regulations; and

Whereas, under the laws of this state, the possessor of land must surround it with a lawful fence, in order to recover damages to crops by cattle; therefore, the farmer or stock man living within or adjacent to a forest reserve, in order to protect his crops and pasturage from forest reserve cattle, must be to the expense of erecting and maintaining a lawful fence; while the forest reserve is not fenced, and is not proposed to be fenced; and the owner of outside cattle must, at his peril, fence or herd his cattle from out of the forest reserve, or pay such price from year to year as shall suit the varying caprice of forest officers; and

Whereas, the making of forest reserves necessitates new and untried procedure, with consequent uncertainties and delays, in order for the settler on unsurveyed land therein to secure the survey of his land, before which he cannot acquire title; and

Whereas, in 1866 Congress passed an act, a portion of which now exists as Section 2477 of the United States Revised Statutes, which grants a free right-of-way across the public lands for the construction of highways, and which enables citizens and local state authorities to speedily construct roads as the necessities of new and quickly growing communities require. (See Okanogan County v. Cheetham, 37 Wash. 682); and

Whereas, said section does not apply to forest reserves; and

Whereas, communities existing on opposite sides of a forest reserve are subject to intolerable delays in attempting to satisfy the requirement of far-distant officers of the government, across such forest reserves, thereby retarding the development and prosperity of the country thus victimized; and

Whereas, it often happens that part of a homesteader's claim, by such indiscriminate withdrawals for forest reserves, is placed partly within and partly without the forest reserve, thereby harassing him with two sets of land laws, a variety of "regulations", and the usual government delays; and

Whereas, if such lands were in private ownership, they would be subjected to state and local taxation, and thereby contribute to the support of government in new states and new communities where revenues are generally inadequate to meet present needs; and the present make-shift provided by Congress as an off-set for the loss of such just revenues from taxation being wholly inadequate and uncertain for the object intended thereby; and

Whereas, it has been the wise policy of the government for more than half a century to grant free water rights upon the public lands; and

Whereas, very recently it has been determined to hamper the acquisition of water rights and rights-of-way within forest reserves, to the great detriment of the general public; and

Whereas, as great protection against the ravages of forest fires can be given outside of forest reserves as within them; and

Whereas, the Congress of the United States, in pursuance of a wise and liberal policy, in 1875 (18 Stat. 482), passed an act granting to railroad companies generally a right-of-way across the public domain, with the rights to take from adjacent land material, earth, stone and timber necessary for the construction of such railroad,etc., which act has aided materially in the upbuilding of the west; and

Whereas, the Act of Congress of March 3, 1899 (30 Stat. 1233) granting a right-of-way for railroads across forest reserves does not grant the right for such railway companies to take material, earth, stone or timber necessary for the construction of such railroad; and

Whereas, the making of a forest reserve segregates the land therein contained from the category of public lands; and therefore said act of March 3, 1875 (18 Stat. 482) does not apply thereto; and

Whereas, the needs of the people require the speedy building of electric railways; and

Whereas, the great source of water power for the operation of such railways, and the providing of electric lights and power for cities and towns, and for the operation of mines, is situated within the limits of forest reserves; and

Whereas, the present Federal restrictions tend materially to defeat the accomplishment of these benign purposes, on account of such forest reserves; and

Whereas, the stock industry is impeded and discouraged by the creation of forest reserves, among other things by the imposition of a pasturage tax and the uncertainty from year to year as to the amount of such tax, and the arbitrary and petty exactions inflicted by forest officer, high and low; and

Whereas, it is the avowed purpose of the forestry service to make the same self-supporting without an appropriation from Congress therefor, and to recruit and organize an army of forestry officers, who must be supported and maintained from the income of the reserve, which means that the communities adjacent to such forest reserves must pay the bill; and

Whereas, more than Seven Hundred Thousand Dollars were wrested from such communities for such purposes during the last year; as against $60,000 the year before; and

Whereas, it is now proposed to increase such exactions in like ratio from year to year; and

Whereas, many hundreds of thousands of acres of land have been placed in forest reserves in two counties in this state, namely, Okanogan county and Ferry county against the unanimous protest of the citizens of these counties, as far as the recent temporary withdrawals are concerned; and

Whereas, the people of a local community can best be trusted to decide what is for their best interests and those of their children and their children's children:

Therefore, we most earnestly and respectfully protest against the making of said temporary withdrawal permanent, and ask that they be immediately set aside and that the lands therein described be at once restored to the public domain.

The Secretary of State is hereby directed to immediately transmit a certified copy of this memorial to His Excellency, the President of the United States of America, to the President of the Senate and Speaker of the House of Representatives of the United States, to the Honorable Secretary of Agriculture, and to each of the Senators and Representatives in Congress, from this State.

Passed by the House, February 27, 1907.

Passed by the Senate, March 6, 1907.

J. A. Falconer, Speaker of the House.

Charles E. Coon, President of the Senate.[18]

THE ROLE OF THE CONSERVATIONISTS

The justification for the attack on the arid West's most important industries was often couched in terms of watershed protection. An embryonic conservationist philosophy—not entirely utilitarian, not entirely religious-aesthetic—dominated the Bureau of Forestry from its earliest days under first Chief Franklin B. Hough (1876), then under Nathaniel Egleston (1883), and then professional forester Bernhard Eduard Fernow (1886), who was firmly in the utilitarian camp. Efforts to pass forest reserve legislation since 1868 had emphasized watershed protection. But again we must remember the conservation movement in the United States began primarily as a result of the wholesale utilization of forests in the northeast for materiel during the Civil War. Northeastern industrial interests who had profited so handsomely by the destruction of forests were annoyed by the incessant clamor of conservationists such as George Perkins Marsh, author of *Man and Nature* (1864), over what industrialists and most citizens viewed as a legitimate utilization of natural resources to win the war.

These same conservationists produced blatant propaganda at times and felt justified in the spread of untruths as long as their objectives were accomplished. As Herbert Kirkland explains about Bernhard Eduard Fernow and early conservationists:

Despite the basically pragmatic nature of the early conservation movement, the leaders were not adverse to playing upon sentiment when it was to their advantage. They realized that the movement had psychological overtones which could only be dealt with through informing the public and influencing public opinion.[19]   A large amount of effort from both the Division of Forestry and the American Forestry Association was, therefore, devoted to propaganda.[20] Fernow might lament the fact that "emotion rather than reason, sentiment rather than argument, are the prime movers of society,"[21] but he also said he "did not wish to deny the importance of sentiment in this whole question of forest reform." Many forestry advocates shared the attitude that "sentiment is the promoter of progress, civilization and manhood" and, therefore, stressed that the development of popular sentiment as applied to the forestry movement should be encouraged as much as possible.[22]

There was also considerable discussion in the latter half of the nineteenth century concerning the effect of forests on general climatic conditions, moisture retention, watershed, and erosion. Perhaps this aspect was overemphasized during this period, as many have claimed; but, besides having a good deal of scientific validity, it also served an important propaganda purpose.[23]

One of the most successful propaganda messages was that forests create rain. The acceptance of this fallacy by Congress led to the passage of the Timber Culture Act of 1873, which encouraged tree planting as a way of drawing rain to the Great Plains.

Historian Ernest Osgood says of the attitude of Congress at this time:

If the arid West did not welcome the farmer as the rich prairie soil had done, it might be induced to do so by Congressional legislation. The thing to do was to adapt the High Plains to the farmer and not the farmer to the High Plains.[24]

The Timber Culture Act and the rain making philosophy behind it were discredited by 1891, but only after the forest reserves had been authorized by Congress.

The anti-grazing, anti-livestock rhetoric did not emerge as a major focus of the preservationist movement until after the forest reserves were authorized. That, incidentally, was after John Muir proved himself a talented lobbyist and gained the attention of the Union Pacific's Harriman and other northern core financial and political interests. For example, anti-livestock harangues were virtually absent from Muir's 1889-90 lobbying for Yosemite National Park. But the entire tenor and pitch of the preservationist message changed dramatically after 1891. Northern core interests, particularly the cattle capitalists and those in the Robert Underwood Johnson circle, wanted to add a publicly acceptable facade to their vicious attack on the future of the western states and the preemptive property rights the states had created on the federal lands through livestock grazing. After 1891 Muir attacked sheepherders with a vengeance—even though he had been one himself—citing livestock grazing as a primary evil.

These attacks focused only on livestock grazing in the western forest reserves. Grazing in the eastern part of the United States was not addressed. Livestock grazing on non-forest reserve federal lands received scant attention. Obviously, the real intent behind these attacks was to defend the forest reserve authority from those most likely to force its repeal.

THE COLONIAL CONSERVATIONISTS

And as the door began closing on the privatization of these lands in the 1890s, western financial interests looked for ways to gain access to timber and grazing on the reserves. The only way appeared to be the establishment of a favored relationship with the Bureau of Forestry. That favored relationship, however, could only be obtained at the price of subjugating the West to the East, by acknowledging the supremacy of eastern capital over western capital, in sum, by renewing the Reconstruction era status of the conquered South and the developing West.

In the late 1890s and early 1900s northern core intent was to command the resources of the forest reserves in much the same way that European nations commanded their foreign colonies. The northeastern United States contained most of the industrial and financial infrastructure of the nation. Northern core politicians—years after Reconstruction—still viewed the conquered South and the developing West primarily as a source of natural resources to maintain that infrastructure.

Historian Richard Bensel noted that following the Civil War, "northern financiers invested heavily in the southern economy and came to control the industrial development and transportation system of the periphery South and West."[25] Of the later period when American imperialism—and the forest reserves—began to flourish, 1905-07, Bensel wrote:

> Unlike the industrial core or its close regional allies in the wheat and corn belts of the Midwest, the South and the mountain West produced raw materials that on the whole sought industrial markets (e.g., cotton and metals) either in the United States or abroad.... In fact the periphery during this period [1905] was composed of debtor regions which imported huge amounts of capital from the northern core....
>
> Because they dominated the national state apparatus, the imperialists of the northern core were not apprehensive about the growing power of central government....
>
> Within the national economy, the northern core was overwhelmingly dominant at the turn of the century, and. . .that dominance was translated into influence over the political economy (military pensions, the tariff, etc.) The periphery resisted, but could not prevent, these developments. However, it sought to prevent a subordinate economic role and political disadvantage from being translated into a permanent feature of the republic. It is this goal that explains periphery hostility to extraordinary preconditions

on admission into the Union for new states and extra-constitutional status for colonial possessions.[26]

As Walter Dean Burnham has said, "The United States was so vast that it had little need of economic colonies abroad; in fact it had two major colonial regions within its own borders, the South and the West."[27]

Maintaining the infrastructure of the Northeast could only be accomplished at the expense of the West and South and beginning in 1905, colonial possessions such as Hawaii, Samoa, Puerto Rico, and the Philippines. The West in particular with its vast reserves of natural resources threatened to eclipse the interests of the established northeastern industrial and financial powers.

The West could attract foreign money. The West could attract wealthy foreign immigrants. Northeastern financial interest understandably wanted this fertile and expanding investment climate for themselves. The attraction of new opportunities in the West constantly drained eastern populations, the source of eastern political superiority over the West in Congress. Eastern Congressmen had tried for decades to quash foreign immigration and foreign money going to the West.

There was no clear policy on just how the resources of the new forest reserves were to be used to create a lasting storehouse of natural resources for the eastern establishment until Gifford Pinchot joined the government and became chief of the Division of Forestry on July 1, 1889, succeeding Fernow.

Pinchot developed the basic blueprint by which nature preservationist views would be downplayed after the reserves were securely and permanently in place. The visions of such people as John Muir embracing the forest reserves as a vast preservationist zone, a vast system of natural cathedrals for nature worshipers, would be forced to give way to the real intention of the architects of the forest reserves. Timber would be harvested in accordance with perceived need and at a rate that would not conflict with the timber industry in the Northeast. In 1890, the northeast was still a major timber producing portion of the nation. New York state had been the leading timber producing state in the nation for a major portion of the 19th century. The Northeast was very sensitive to any competition from new timber producing areas. When H.R. 7901 was being debated in the 50th Congress in 1888, New York State was pushing a resolution through Congress supporting a tariff on imported lumber products from Canada. The policies of Gifford Pinchot would not create unacceptable competition for the northeastern lumber interests.

Preservationist attacks on the rancher's range rights in the forest reserves, particularly the articles by Robert Underwood Johnson and John Muir, had obviously been successful—perhaps to the point of overkill. The 1894 ban on grazing not only threatened a very significant portion of the west's economy, but also asserted federal control over state water and the private water rights of individuals. Forest Reserve administrators and defenders grew alarmed at the ferocity of the western protests.

The bitter western response to the forest reserves during the latter part of the Cleveland administration forced forest reserve proponents to relinquish their anti-grazing stance. They feared that a continued ban on grazing "might provoke total flaunting of the law, disrespect for the reserves, and strong political pressure for the return of the reserves to the public domain." Underlying this political pressure was the specter of the ranchers' pre-existing rights and the U. S. Supreme Court's validation of those rights in *Buford* v. *Houtz*.

Forest reserve administrators decided it was better to permit grazing in the reserves than to chance loosing the reserves altogether. The effort to usurp the states' ownership of non-navigable waters in the forest reserves by prohibiting the use of those waters threatened to raise issues which could lead to the repeal of Section 24 and the elimination of the entire forest reserve system. As historians Paul Gates and Robert Swenson stated, the Department of the Interior and later the Forest Service would deem it not only advisable to permit grazing, but also politically expedient for the integrity and expansion of the new government forest lands.[28]

Chapter Six Footnotes

1. *Buford* v. *Houtz*, 133 U.S. 320, 326-328 (1890).
2. Robert Underwood Johnson, *Remembered Yesterdays*, Little, Brown and Company, Boston, 1923, p. 288.
3. Wolfe, *Son of the Wilderness: The Life of John Muir*, p. 271.
4. The order is described in Frederick V. Coville, *Forest Growth and Sheep Grazing in the Cascade Mountains of Oregon*, U. S. Department of Agriculture, Division of Forestry Bulletin no. 15, p. 10.
5. Ise, *The United States Forest Policy*, p. 121.
6. U. S. Congress, *Senate Report of the Committee Appointed by the National Academy of Sciences upon the Inauguration of a Forest Policy for the Forested Lands of the United States*, 55th Cong., 1st Sess., 1897, S. Doc. 105, vol. 5, p. 7.
7. Steen, *The U.S. Forest Service, A History*, pp. 30-31.

8. Fernow said the proclamation had "stirred up such an antagonism as we have never had before." Fernow to H. G. deLotbiniere, March 6, 1897, RG 95-2, RFS; Congressman John Bell of Colorado recommended the dissolution of the reserves according to G. Michael McCarthy, *Hour of Trial: The Conservation Conflict in Colorado and the West 1891-1907*, University of Oklahoma, Norman, 1977, p. 55.

9. Lamm and McCarthy, *The Angry West: A Vulnerable Land and Its Future.*

10. As quoted in Samuel T. Dana and Sally K. Fairfax, *Forest and Range Policy: Its Development in the United States.*, 2d ed., McGraw-Hill, New York, p. 58.

11. Rowley, *U.S. Forest Service Grazing and Rangelands: A History*, p. 30.

12. Ibid. p. 31.

13. "Description of Township 16 S., R 31 E, W M in Grant County, Oregon," RG 95, Entry 13 (43), Vol H, National Archives.

14. "Description of Township 15 South, Range 26 East, W. M. in Grant County, Oregon," RG 95, Entry 13, (42) Vol H, National Archives.

15. "Protest Against Forest Reserve," RG 95, Entry 13, (12) Vol H, National Archives.

16. "Description of Township 10 South, Range 33 E. W. M. in Grant County, Oregon," RG 95, Entry 13, (45) Vol H, National Archives.

17. "To the editor" [of an unidentified Oregon newspaper], RG 95, Entry 13, (47) Vol H, National Archives.

18. "[Washington State] House Memorial No. 6 [to the President, his Cabinet, and Congress]," signed by J. A. Falconer, Speaker of the House, and Charles E. Coon, President of the Senate, RG 95, Entry 13, (7) Vol H, National Archives.

19. Kirkland here cites *Proceedings of the American Forestry Association*, XI, 162.

20. Kirkland here cites *Ibid.*, p. 35; Bernhard Eduard Fernow, *Report upon the Forestry Investigation of the United States Department of Agriculture, 1877-1898*, Government Printing Office, Washington, D. C., 1899, p. 24.

21. Kirkland here cites Bernhard Eduard Fernow, *Economics of Forestry*, Thomas Y. Crowell and Company, New York, 1902, p. 9.

22. Kirkland here cites "The Forestry Meeting at Lancaster," *Forest Leaves*, III (April, 1890),

23. Kirkland, *The American Forests, 1864-1898: A Trend Toward Conservation*, pp. 134-135.

24. Ernest Staples Osgood, *The Day of the Cattleman*, The University of Chicago Press, Chicago, 1929, pp. 194-195.

25. Richard Franklin Bensel, *Sectionalism and American Political Development: 1880-1980*, p. 73.

26. Ibid., pp. 92-93.

27. Burnham, "The End of American Party Politics," p. 16.

28. Gates and Swenson, *The History of Pubic Land Law Development*, p. 569.

7

THE WEST ANSWERS

Gifford Pinchot, who became known in later years as the father of the Forest Service, referred to the grazing question as "far and away the bitterest issue of the time."[1] Forest reserve administrators in the late 1890s realized that if they were to save the reserves at all they must make some accommodation for grazing. This posed two immediate questions. If livestock grazing in the reserves must be accepted, how could it be done without reinforcing preemptive private property rights? If the federal government was to gain control of the non-navigable waters on the federal lands, how could use of that water according to *local law and custom and the rules of the court* be allowed without reinforcing the states' rights to control of the water?

Initially forest reserve officers tried to circumvent vested preemptive rights by suggesting that grazing privileges would be granted to *new* livestock interests at the expense of the established livestock operator. New operators could make no valid claim to preemptive property rights. Administrators were attempting to quiet the grazing controversy by replacing those graziers who claimed property rights in the reserves with a loyal clientele of stockmen who owed their grazing privileges to the government. This was the policy of the northeastern capitalists. It was an attempt to disenfranchise the early prior appropriators in favor of northeastern interests who expected to enjoy a government-protected monopoly in the forest reserves.

Residents of Grant County, Oregon, expressed their opposition to this policy in their "Protest and Remonstrance to the President of the United States" over proposed forest reserve withdrawals in 1902:

> [T]he creation of the said proposed forest reserve is supported by advocates of leasing of government land in hope of securing permits for grazing thousands of head of stock on the said reserve to the detriment of resident stockmen of Grant County.[2]

That newcomers were anxious to usurp the rights of established settlers is further illustrated in a letter to Binger Hermann of the General Land Office, Washington D. C. dated December 14, 1899 from J. A. E. Psujdam of Clifton, Arizona. While the writer was certainly not one of the northeastern capitalists and there is no indication that he had any direct connection to them, his correspondence does illustrate the pervasive anti-western, anti-southern climate of opinion during this period. The letter, though poorly drafted with numerous misspellings, succinctly expresses the usurpers' viewpoint. The writer is apparently trying to ingratiate himself with forest reserve officials. He would like to see established cowmen removed from the reserves. He emphasizes the

North-South conflict and his loyalty to northern interests. He ends his letter with an appeal for watershed protection. I quote the letter here in its entirety, exactly as written.

> Your note of November the 10 at hand. I don't think you quite under Stand my Diagram that I sent you. this Strip of Land that I refer to being 80 by 40 in Yocham Co. is part of the Black Mesa timber resurv. What I was finding fault with was that they had Cut this Strip of Land in two and Leaving the timber that was the most Conveniant to get at out of the resurv. they have left the Pastures all So. I was over in N. Mexico a Short time agoe and one of the Sign Riders was there Doeind his duty and I heard Some of the Cow men say that they had out to get to gather and run him out. I Being a Soldier of the 60 I advised them to let that Part out but I under Stand that they abused him terbily. Now this is why I Say that those Parties that are on the resurv and have it under Pasture Should move. for in one cence of the word they are not Citizens of the union. they have Sprund from the ignorant Class of the South and Would take up arms a gainst the union Just the Same as there fathers Did in 1861. I have Talked With Lots of them and they have no use for a yankie. I was a Soldier in the Civil War I Dun my duty as a Soldier the government paid me 16 Per month I was with Phil Sheriden and John A. Logan and was wounded twice but I am all rite the government Dont owe me any thing. But when I no that these Kind of People fences up the Publick Domain then i am at a Loss to no what I was there for.
>
> When they are moved from the Publick Domain the water will increas. in a few years But if they are Left it will Decreas.
> Yours Truly
> JAE Psujdam
> Box 172 Clifton Ariz.[3]

The Washington State legislature complained against allowing the livestock of newcomers (referred to as forest reserve cattle) into the new forest reserves at the expense of the established settler and stockman.

> [The] stockman living within or adjacent to a forest reserve, in order to protect his crops and pasturage from forest reserve cattle must be to the expense of erecting and maintaining a lawful fence; while the forest reserve is not fenced, and is not proposed to be fenced; and the owner of outside cattle [settlers' cattle grazing adjacent to but not on the reserve] must, at his peril, fence or herd his cattle from out the forest reserves, or pay such price from year to year as shall suit the varying caprice of forest officers . . . .[4]

Northern core interests did everything they could to gain control over the forest reserves. This effort is well illustrated in the following 1900 letter from E. H. Libby, president and general manager of a Washington water and power company, to James Glendenning, U. S. Forestry Superintendent, Lewiston, Idaho. The late 1800s and the turn of the century saw tremendous growth in the amazing new power technology called electricity. The watersheds of the western mountain ranges contained the nation's best hydroelectric generating sites. Control of these sites was a prime objective of northeastern industry and capital. The preemptive property right claims of established western stockmen needed to be neutralized if northeastern control of these sites was to be achieved.

Sir,
Will you be so good as to inform us as to the necessary steps to be taken to secure an examination of the Blue Mountain at the headwaters of the Asotin Creek and other streams rising in the vicinity? Last year the water in Asotin Creek was much lower than ever before in the memory of the "oldest inhabitant", and so low as to greatly endanger important interests dependent upon the uniformity of the water supply. There are other important streams which rise in the same neighborhood, upon the waters of which the prosperity of several valleys is dependent, especially the Grand Ronde, George Creek, Alpowa, Tucannon, Touchet, Walla Walla, etc. The timber on the Blue Mountains is much scattered and settlers are cutting it off rapidly, although the soil is thin and of little value except for hay. Other large areas have been burned off, as I saw last Summer during four days spent in riding around in the vicinity of the heads of Asotin and George creeks. That some steps should be taken looking toward the setting aprt [sic] of a forestry reserve in the Blue Mountains admits of no doubt in my opinion, and to this end I beg that you will kindly advise as to the procedure required to secure the examination preliminary to such setting apart of a reservation.
Respectfully,[5]

The letter at first appears to be an expression of concern by a western interest justifying the establishment of forest reserves to protect watershed values. The letter also blames settlers producing hay (stockmen) for damaging the environment which allegedly has resulted in a low water runoff year. The letter as written appears to be an acknowledgement by western interests that conservationist and northern core concerns were correct about the western resource lands. However, doubt is cast on the actual intent of the letter when one reads a list of the Board of Trustees on the letterhead and where they reside:

Charles Francis Adams,	Boston
G. St. L. Abbott,	Boston
Wm. H. Boker,	Boston
William Whittaker,	Boston, Consulting Engineer
George W. Bailey,	Boston, Counsel
Elbert Wheeler,	Boston, Treasurer
E. H. Libby, President and General Manager	

As historian G. Michael McCarthy and Colorado Governor Richard Lamm wrote in *The Angry West*,

Throughout most of the century, huge politico-economic combines—many with eastern ties—dominated the critical western water-power industry. By 1914, some twenty-eight corporations owned 90 percent of the water power developed on western public lands—and six corporations owned 56 percent of that. The Utah Securities Corporation, dominated by outsiders, owned 30 percent in Colorado, 70 percent in Utah, and 20 percent in Idaho. Montana Power owned 97 percent of the power development in that state, and as late as 1970, Montanans held only 36 percent of the company's stock.[6]

Proponents of the forest reserves in the Northeast were quick to employ the mass media in support of their attack on the western states. Northeastern forestry associations, such as the Maine State Forestry Association, were active from 1891 producing material to feed the media in the promotion of forest reserves. The American Forestry Association, headquartered in Washington, D. C., was a major promoter of the forest reserves. Such periodicals as *Garden and Forest*, *Century*, and the *Atlantic Monthly* magazine, all published in New York City, promoted the attack on the western states' resource base. The *New York Times* and *Saturday Evening Post* were tools of the eastern capitalists and regularly told the general public that the subjugation of the western states by northern core political and financial interests—in the guise of "conservation"—was in harmony with the *public good*.

Western stockmen replied to this attack on their property rights in the new forest reserves by citing the Supreme Court's 1890 ruling in *Buford* v. *Houtz*. They said that the Congress of the United States had granted an implied license to graze on unclaimed public lands.

Reserve defenders did not listen. They claimed that Section 24 of the 1891 Act removed these forest reserves from the public domain, thus nullifying the stockmens' claims. They asserted that Section 2 of the 1891 Act had repealed the preemption laws, further nullifying the stockmen's claims to vested preemptive rights in the land.

Western states and livestock interests countered that the Forest Reserve Act and repeal of the preemption laws could not extinguish rights established prior to 1891. Non-navigable waters, even though arising on federal lands in the forest reserves were property of the respective states, they argued. Under the prior appropriation doctrine of water rights, which dominated many of the western states, the first party to put the water to beneficial use could obtain private title to the water. Those stockmen who had obtained title to the waters on western land had control of the land, especially in the arid areas where water was limited. Many forest reserves were being created in arid areas where no commercial timber was present.

Ownership of the water by the stockmen presented an unsolvable dilemma to forest reserve proponents. The stockman did in fact control the land. The new forest reserve lands were in fact split estate lands.

THE PRIOR APPROPRIATION DOCTRINE

At the heart of the legal doctrine of prior appropriation is the concept of *possessory interest* or *possessory rights*. It was this doctrine of *prior appropriation* or *preemption*, having its roots in ancient law, which had been a major factor in the privatization of the public lands of the United States from the beginning of the nation. It was this doctrine of prior appropriation or preemption which was being practiced on the western rangelands prior to the wholesale departure from traditional land law in the United States which came about with the addition of Section 24 to the Forest Reserve Act of 1891.*

A 1933 report released by the Department of Agriculture spoke of the establishment of forest reserves as "a bold departure."

> This action flew in the face of accepted ideas regarding the functions of the Federal Government. It involved a reversal of the course which had been pursued with relation to the public domain from the foundation of the Union. It conflicted with the whole American tradition of individualism and preference for private property ownership.[7]

The doctrine of appropriation and prior use was a valid method of acquiring private property out of the public domain. The western livestockmen took their cue for privatizing the grazing lands from the farming and mining settlers in their midst. The miner and the farmer appropriated water and land and the U.S. Government recognized title to both.

> The prevailing Western Doctrine of prior appropriation as it is now recognized throughout the 17 contiguous western states and Alaska, is traceable chiefly to local customs and regulation developed spontaneously on public land.... With considerable uniformity, these simple but effective principles became formalized into legal doctrine by decision of courts and enactments of legislatures. Upon this foundation have been built the current complicated and voluminous water codes and case laws of the West.[8]

The seeds of the appropriation doctrine are discernable in the statutes of three general movements of great historical and economic importance: the Spanish settlements in parts of the southwest, the Mormon colonization of Utah, and the California gold rush.[9]

In 1898, the Territorial Supreme Court of New Mexico said: "The Law of prior appropriation existed under the Mexican republic at the time of the acquisition of New Mexico, and one of the first acts of this government was to declare that "the laws hereto in force concerning water courses... shall continue in force."[10]

The appropriation doctrine, as it applied to New Mexico and Arizona, is thought by some scholars to have its roots in ancient Rome. Water authority Wells A. Hutchins remarked, "According to one school of thought, the Spanish settlers brought this doctrine from Europe with their civil law, which had been derived from the civil law of Rome."[11]

*Supra, Chapters 3 and 5.

Utah water law began when Brigham Young brought the Mormons into the Great Salt Lake Valley in 1847. Even though the area technically belonged to Mexico, as a practical matter its land grants and water law had not been extended there. The Church of Jesus Christ of Latter Day Saints took effective possession of all the country extending to the California border, ignored Mexican and U. S. laws, and went about making its own legal system.[12] The saints laid out townsites and farms and allotted specific parcels to members of the church. Without recourse to or interference from the United States Government, the Mormons improvised a temporary system of land titles, pending the acquisition of definitive federal grants, and the roots of a permanent system of water titles. When the Mormons formally established the State of Deseret on July 3, 1849, these early possessory titles were recognized by Deseret law.[13]

From the earliest days the Mormon Church approved and supervised the group diversion of water for beneficial use, much as the California Gold Rush miners did, by custom and local rules. Early legislation made grants of water privileges, authorized the making of grants, and vested in the county courts control over appropriation of water.[14]

Utah Territory passed a law in 1880 recognizing accrued rights to water acquired by appropriation or adverse use, but did not contain a specific authorization to appropriate water.[15] Nevertheless, appropriation was widely practiced by custom in the settled parts of Utah despite the absence of authority from either the legislature or the supreme court. Shortly after entering the Union in 1896 as the forty-fifth state, Utah authorized the prior appropriation doctrine as the law of the land.[16]

California and Nevada water law was primarily influenced by the California gold rush and the customs of the miners. In the words of the United States Supreme Court, speaking through Justice Stephen Johnson Field who had been Chief Justice of California, the miners "were emphatically the law-makers, as respects mining, upon the public lands in the State." The principles embodied in their customs were the foundation of later law:

> In every district which they occupied they framed certain rules for their government, by which the extent of ground they could severally hold for mining was designated, their possessory right to such ground secured and enforced, and contests between them either avoided or determined. These rules bore a marked similarity, varying in the several districts only according to the extent and character of the mines; distinct provisions being made for different kinds of mining, such as placer mining, quartz mining, and mining in drifts or tunnels. They all recognized discovery, followed by appropriation, as the foundation of the possessor's title, and development by working as the condition of its retention. And they were so framed as to secure to all comers, within practicable limits, absolute equality of right and privilege in working the mines. Nothing but such equality would have been tolerated by the miners, who were emphatically the law-makers, as respects mining, upon the public lands in the State. The first appropriator was everywhere held to have, within certain well-defined limits, a better right

than others to the claims taken up; and in all controversies, except as against the government, he was regarded as the original owner, from whom title was to be traced. But the mines could not be worked without water. Without water the gold would remain forever buried in the earth or rock. To carry water to mining localities, when they were not on the banks of a stream or lake, became, therefore, an important and necessary business in carrying on mining. Here, also, the first appropriator of water to be conveyed to such localities for mining or other beneficial purposes, was recognized as having, to the extent of actual use, the better right. The doctrines of the common law respecting the rights of riparian owners were not considered as applicable, or only in a very limited degree, to the condition of miners in the mountains. The waters of rivers and lakes were consequently carried great distances in ditches and flumes, constructed with vast labor and enormous expenditures of money, along the sides of mountains and through cañons and ravines, to supply communities engaged in mining, as well as for agriculturists and ordinary consumption. Numerous regulations were adopted, or assumed to exist, from their obvious justness, for the security of these ditches and flumes, and the protection of rights to water, not only between different appropriators, but between them and the holders of mining claims. These regulations and customs were appealed to in controversies in the State courts, and received their sanction; and properties to the values of many millions rested upon them.[17]

Historians Charles H. Shinn in 1885 and William E. Colby in 1935 both traced western water law to its roots among the Germanic peoples of the Old World: "Certain it is that the 'Forty Niners' came to California from many countries. They may well have brought with them some knowledge of the old Germanic customs and applied this knowledge in their new environment."[18]

Hutchins says of the development of possessory rights:

The appropriation doctrine developed chiefly on the public domain. For years the owner of these lands—The Federal Government—made no move either to assert or to grant away its water rights. The miners were trespassers, and so their claims to use of the water were not good as against the Government. However, in the absence of specific State or Federal legislation authorizing the appropriation of water, the customs established in the mining camps of recognizing rights to the use of water by appropriation—"first in time, first in right"—eventually became valid local law. This came about because of the policy of the courts to recognize miners' claims as possessory rights that were good among themselves and as against any other claimant but the Government.[19]

Because the statutes were silent, common law became an underpining of the doctrine of possessory rights. Without direct precedent or specific legislation, the courts had to resort to common law analogies, which is uncomfortable ground for judges. For example, in 1857 the California Supreme Court complained that the California judiciary had been burdened with responsibilities not faced by other American courts in mining cases.[20]

Fortunately, the common law analogies were not too far-fetched. In one case relating to controversies over possession of land between persons without title in which the real owner was absent, the analogy was made to the rules of law regarding priority of possession of land. "The diversion of water was declared to be the equivalent of possession and the doctrine was laid down that he who was first in time was first in right."[21]

The high court resorted to another common law principle in an 1856 case related to the doctrine of presumption, under which it was presumed from the absence of specific legislation that everyone who wished to appropriate water or to dig gold on the public domain within California had a license to do so, provided that the prior rights of others were not thereby infringed.[22] The miners had certainly not kept their activities secret and Congress did nothing about it one way or another, so the presumption was completely reasonable.

An extensive body of case law derived from common law grew up protecting possessory interests. Mining and diverting water were ruled to be equally conferred and stood on an equal footing.[23] The right to the use of running water on the public domain was held to exist without private ownership of the soil, on the basis either of prior location on the land or of prior appropriation and use of the water, a ruling that permitted cattlemen to survive.[24] Conflicts between land claimants and water claimants were to be decided by the fact of priority in time of either land location or water diversion.[25] Gradually the whole doctrine of prior appropriation and beneficial use evolved, splitting various interests out of the federal estate.

Possessory interest through prior appropriation applied to land as well as water. The priority principle was applied in early California cases as between appropriators of water for mining purposes, and was soon extended to other purposes as well.[26] The courts had to authorize this extension, for in the mining areas it was argued that a prior appropriation could be made solely for the purpose of mining and there was as yet no Federal or State legislation on the subject. In one case, the right to use the water of a stream was claimed by a prior appropriator for operating a sawmill and by a subsequent appropriator for working mines.[27] The upstream mining diversions from a water supply insufficient for both claimants prevented operation of the mill for 5 months of the year. The California Supreme Court affirmed the judgment of the trial court in issuing an injunction against the miners. The decision stated that under state policy the prior appropriation of either land or water on the public domain entitled the holder to protection in its quiet enjoyment.

At first these claims on land and water in the public domain were viewed as invalid against the federal government. The claimants were technically trespassers on federal lands. Later these original claims to land and water were recognized and the government conveyed title accordingly. Thus the doctrine of appropriation developed at first with the silent acquiescence of the government, and then with congressional recognition. The Pacific States and Territories enacted legislation which was supported by the courts in which priority gives the better right.

The appropriation doctrine had been accepted throughout the West by 1875. This was the period during which stockmen—both original settlers and usurping capitalists—had competed fiercely to lay claim to the western grazing lands. Their claiming of the waters on federal lands gave them control of grazing lands within the distance livestock would normally travel to and from water. It was this competition which led to much of the overgrazing which conservationists were later to magnify disproportionately in their attacks on the western livestock interests. About this period Hutchins says:

> During the first 25-year period following the Gold Rush —approximately 1850 to 1876—the appropriation doctrine was adopted by State or Territorial statute, or was recognized by high court decision, or both, in Arizona, California, Colorado, Montana, Nevada, New Mexico, and Wyoming.[28]

WITHDRAWALS

Throughout the seventies and the eighties Western stockmen became acutely aware of the potential for restrictive land withdrawals from the public domain. In 1864 the federal government conveyed Yosemite Valley—but not the surrounding high Sierra territory in today's national park—to California for a state park. Livestock grazing, timber cutting, and mining were excluded.

In 1890 the high Sierra country surrounding Yosemite Valley—but not the valley itself—was made a national park primarily through the efforts of two men: Robert Underwood Johnson and John Muir. Johnson was the editor of *Century* magazine and saw himself as a leader in "political, religious, artistic and social opinion."[29]

In 1889 Johnson met Muir and recruited his talents to establish a national park at Yosemite.* Muir was an eccentric genius who had spent years in the Yosemite area as a sheepherder, guide, and sawmill operator. He had emigrated from Scotland as a child. His memory of large sheep corporations disenfranchising the small sheep owners of their grazing lands in his native Scotland left him with little sympathy for large corporate interests, especially if they were sheep graziers.[30] One of the major roadblocks to Johnson's objective was the sheepmen who could claim a prior appropriation right to the grazing in Yosemite Valley. The herds of sheep entered Yosemite Valley each summer. Muir had often herded those very sheep.

Although Muir's first significant national recognition had been based on his study of glaciation as the probable force in creating the Yosemite Valley, his knowledge of sheep and his bitter memories of corporate sheep graziers in his native Scotland made him an ideal candidate to help promote anti-grazing rhetoric in support of the Yosemite area for a park. Johnson wrote of his offer in recruiting Muir:

*Cf. supra, p. 116.

I told him that if he would agree to write for the *Century* two articles,— the first on "The Treasures of the Yosemite", to attract general attention, and the second on "The Proposed Yosemite National Park", which he and I should propose, and the boundaries of which he should outline,—I would ask Mr. Gilder by telegraph on our return to the Valley if I might engage the articles. We would illustrate them with pictures of the wonderful natural features of the Goverment lands proposed to be taken for the park, and with these pictures and the proofs of Muir's articles I would go to Congress (where I was to be much engaged in the international copyright campaign), and advocate its establishment before the committees on Public Lands. One of the members of the House Committee was Judge Holman of Indiana, whose circuit had adjoined that of my father when he was Judge of Court. The two, though political opponents,—Holman being a Democrat and my father a radical Republican,—had nevertheless been friends, and I felt sure that Holman would be predisposed to the scheme. I also knew Mr. Plumb of Kansas, who was chairman of the Senate Committee, and I believed he would help. I told Muir that I thought there would be no serious objection to such a measure and that in my judgment it would go through.[31]

Muir was ideally suited to the objectives of park proponents such as Johnson. In an era that thrived on legends and heroes, John Muir became the Kit Carson or Buffalo Bill of the conservation movement. Muir had charisma. A brilliant man despite his peculiarities, he could speak and write in glowing poetic phrases about the wonders of the natural world. The media and the public seized on his eccentric charismatic qualities and further built his image by comparing him with Elijah, Jeremiah, John the Baptist, and even Christ.

Muir was encouraged to speak publicly and write, although the attempts to civilize this self-made man of nature were often difficult. In just one year, however, the strategy worked. The creation of the park was accomplished in 1890. Muir's well-publicized paeans to the "glory of nature" helped significantly to drown out the objections of western commercial interests. Johnson had established Muir as a useful spokesman of the conservation movement, a development which was not lost on other national conservationist leaders in the nation's northeastern quadrant.

A sidelight of the Yosemite campaign has been lost to those who regard Muir as a folk hero: He was allied with and in the pay of railroad magnate Edward H. Harriman. As Johnson wrote of the 1904-05 campaign to convince California to retrocede the Valley of Yosemite back to the federal government to make a complete management unit, "I must not dismiss this topic without saying that California and the country are much indebted for the success of this measure of retrocession to Edward H. Harriman, President of the Union Pacific Railway." Muir was "a warm friend of his," and wrote to him asking for lobbying assistance. Harriman, "having a large acquaintance in the legislature...was able to be of much use to us in the final contest." Johnson did not mention that Harriman also obtained a monopoly for the railroad route to Yosemite.

When the other areas of the West were proposed for forest reserve status after 1891, John Muir was called upon again. He proved his worth once more. Grazing, lumbering, mining, and all forms of trespass were excluded from the early forest reserves, sending a further ominous message to western settlers and the western states.

John Muir's career as a leading spokesman for the preservationist arm of the conservation movement was now well launched. Immediately he was courted and financed by other major northeastern financial interests besides Harriman who wanted to lock up the resources of the West. He was feted in New York and introduced to Harvard scientists including eminent dendrologist Charles S. Sargent and biologist Alexander Agassiz who would in 1897 successfully recommend the lockup of 21 million acres in the famous "Washington's birthday" forest reserves. Muir would be the spokesman for the northern core financial and industrial interests as well as the preservationists—at least until his usefulness was over.[32]

WATER LAW EXPANDS

The development of water law in the West reflected this growing concern with the potential of federal intrusion on states' rights. The second period—1875 to 1900—saw local statutory recognition of the appropriation doctrine spread through all Western States and Territories. Legislatures authorized the prior appropriation of water for beneficial purposes in each of the present jurisdictions of Idaho, Kansas, Nebraska, North Dakota, Oklahoma, Oregon, South Dakota, Texas, Utah, and Washington. In many of these jurisdictions water had been used for mining and agricultural purposes for many years.[33]

How did the federal government react to this trespass of settlers upon the public land? How did these "properties" established by "local law and custom and court decisions" become recognized as real property by the United States? How did the farmer and miner acquire fee title to their possessory interests on the public lands? Why did the stockman believe he would be treated the same in his possessory interests? Hutchins details the development of the law:

[T]he United States was the owner of the lands upon which the American customs of appropriating water for mining purposes originated, and on which lands these customs were practiced in the early development of the appropriation doctrine. The significance of this fact in its impact upon the establishment of the doctrine in the West calls for strong emphasis.
Period of silent acquiescence.—After the discovery of gold, diversions of water on the public domain for mining and other purposes were made for years before Congress took direct notice. Possessory titles to land and water representing in the aggregate great wealth were acquired and conveyed from one holder to another, with the sanction of the courts, on the assumption that the silence of Congress indicated tacit consent.[34]

Shortly after the close of the Civil War in 1865, proposals were made in Congress that the Government withdraw the mines on the public domain from the miners, and operate and sell them in order to obtain revenue to help pay the war debt. Western Senators and Representatives thereupon made a forceful and successful campaign to halt this movement, the culmination of which was the enactment on July 26, 1866, of an act expressly confirming the rights of miners and appropriators that had been recognized only tacitly theretofore. . . .[35]

Act of 1866.—. . . What it did was to take cognizance of the customs and usages that had grown up on the public lands under State and Territorial sanction and to make compliance therewith essential to enjoyment of the Federal grant.

According to the United States Supreme Court, this Congressional act was more than the establishment of a rule for the future.[36] It also constituted recognition of a *preexisting right. . . .*

The section of the Act of 1866 confirming such rights was, in the Court's opinion, "*rather a voluntary recognition of a pre-existing right of possession, constituting a valid claim to its continued use, than the establishment of a new one.*"[37] [My italics.] The act merely recognized the obligation of the Government to respect private rights which had grown up under its tacit consent and approval.[38] It proposed no new system, but sanctioned, regulated, and confirmed the system already established, to which the people were attached.

Act of 1870.—An amendment of the Act of 1866, enacted July 9, 1870, provided that all patents granted, or preemption or homestead rights allowed, should be subject to any vested water rights, or rights to ditches and reservoirs used in connection therewith, as may have been acquired under or recognized by section 9 of the Act of 1866.[39]

The Act of 1866 had recognized water rights and rights of way on public lands as against the Government. The amendment of 1870 clarified the intent of Congress that the water rights and rights of way to which the 1866 legislation related were effective not only as against the United States, but also against its grantees—that anyone who acquired title to public lands took such title burdened with any easements for water rights or rights of way that had been previously acquired, with the Government's consent, against such lands while they were in public ownership. . . .

Desert Land Act of 1877.—This act provided that water rights on tracts of desert land should depend upon bona fide prior appropriation; and that all surplus water over and above actual appropriation and necessary use, together with the water of all lakes, rivers, and other sources of water upon the public lands and not navigable, should be held free for appropriation by the public for irrigation, mining, and manufacturing purposes, subject to existing rights.[40] This act applied specifically to Arizona, California, Idaho, Montana, Nevada, New Mexico, North Dakota, Oregon, South Dakota, Utah, Washington, and Wyoming. An amendment in 1891 extended the provisions to Colorado.[41]

The question whether the desert land legislation was limited to desert lands was not decided by the United States Supreme Court until 1935. . . .

The question was settled by the United States Supreme Court in 1935 in the California Oregon Power Company case, wherein it was held that the Desert Land Act applied to all the public domain in the States and Territories named, and that it severed the water from the public lands and left the unappropriated waters of nonnavigable sources open to appropriation by the public under the laws of the several States and Territories. . . .[42]

However, in view of the fact that so much land in the West was in public ownership during the period under consideration, the Congressional legislation was a powerful factor in the spread of the appropriative principle throughout the West.[43]

Are these water rights taken up under the appropriative doctrine real property rights? Hutchins writes:

—The appropriative right is a right of private property. It is subject to ownership, disposition, and litigation as in the case of other forms of private property.
—The appropriative right is valuable property.
—The general rule in the West is that the appropriative right is real property. . . .

In general.—The appropriative right is an interest in real estate. [Although an interest in realty, the appropriative right is a right of use and is subject to loss as a result of nonuse. It thus differs from title to land.] As a general practical matter, a person who is legally competent to own title to land in a particular jurisdiction has equal competence to hold title to an appropriative water right therein.[44]

If the appropriative right to water is a right of private property, subject to ownership disposition and litigation as in the case of other forms of private property, then it is easy to see how the stockmen of the late 1800s believed they had a valid property right claim to the grazing land serviced by the water. After all, it was the grazing that constituted proof of beneficial use. Without recognition of grazing as a beneficial use the claims to the water could not be valid, yet the states had clearly recognized grazing as a beneficial use both by custom and statute.

Opponents of grazing rights said that grazing on the forest reserves was a privilege granted by the government after the establishment of the first grazing rules in the late 1800s and early 1900s. They argued that grazing permits had no basis in pre-existing rights. But, as we shall see, the grazing permit system developed by the government in the forest reserves—which became today's national forests—and later on the remaining federal lands, was in fact a recognition of those original stock water rights. Water rights which had been legally recognized by both state and federal law as valid by virtue of livestock grazing which had occurred prior to the 1891 Act that created the forest reserves and repealed preemption.

The U.S. Supreme Court was explicit in explaining that the Act of July 26, 1866, was in fact addressing the recognition of pre-existing rights, *"Rather a voluntary recognition of a pre-existing right of possession, constituting a valid claim to its continued use, than the establishment of a new one."*[45] [My italics.]

Chapter Seven Footnotes

1. Pinchot, *Breaking New Ground*, p. 45.

2. People of Grant County, Oregon, *Protest and Remonstrance: The People of Grant County, Oregon, Object to More Than Half of the County's Area Being Withdrawn from Settlement*, Forest Service Field Records, 1898-1904, R. G. 95, National Archives.

3. J. A. E. Psujdam to Binger Hermann, December 14, 1899. Forest Service Field Records, 1898-1904, E 13 (40-41), R. G. 95 National Archives.

4. Memorial from Washington State Legislature, Forest Service Field Records, 1898-1904. R. G. 95 National Archives.

5. Letter from E. H. Libby, president of a water and power company, to James Glendenning, U.S. Forestry Superintendent, Lewiston, Idaho. Forest Service Field Records, 1898-1904. R. G. 95 National Archives.

6. Lamm and McCarthy, *The Angry West: A Vulnerable Land and Its Future*, p. 14.

7. U.S. Department of Agriculture, *National Plan for American Forestry*, I, Washington, D. C., 1933, p. 746.

8. Hutchins, *Water Rights Laws in the Nineteen Western States*, p. 159.

9. The three paths to prior appropriation can be traced most directly through: Southwest: Report and Recommendations of Committee of National Reclamation Association, "Preservation of Integrity of State Water Laws," (1943); Mormons: Edward Jones Allen, *United Order Among the Mormons*, Studies in History, Economics and Public Science #419, Columbia University, New York, 1936; and California Gold Rush: John Robert Umbeck, *A Theory of Property Rights, With Application to the California Gold Rush*, Iowa State University, Ames, 1981.

10. *United States* v. *Rio Grande Dam & Irr. Co.*, 9 N. Mex. 292, 306-307, 51 Pac. 674 (1898), reversed on other points, 174 U. S. 690 (1899).

11. Hutchins, *Water Rights Laws in the Nineteen Western States*, vol. 1, p. 162. Roman water law definitely contained provisions resembling both prior appropriation and beneficial use doctrines. However, the transmission of Roman law to Mexico is not supported in either the laws of Spain or of Mexico, which are silent on such provisions. See Eugene F. Ware, *Roman Water Law, translated from The Pandects of Justinian, containing all of the Roman law, concerning fresh water, found in the Corpus Juris Civilis*. West Publishing Company, St. Paul, 1905. p. 83.

12. Allen, *United Order Among the Mormons*, p. 94.

13. Laws and Ordinances of the State of Deseret, Compilation 1851, Salt Lake City, 1919.

14. George Thomas, *Early Irrigation in the Western States*, University of Utah, Salt Lake City, 1948.

15. Utah Laws 1880, ch. XX.

16. Utah Laws 1897, p. 219.

17. *Jennison* v. *Kirk*, 98 U. S. 453, 457-458 (1879).

18. See Charles Howard Shinn, *Mining Camps, A Study in American Frontier Government*, Charles Scribner's Sons, New York, 1885; reprint Knopf, New York, 1947, pp. 11-35; and William E. Colby, "The Freedom of the Miner and Its Influence on Water Law," published in *Legal Essays, in Tribute to Orrin Kipp McMurray*, 1935, pp. 67-84.

19. Hutchins, *Water Rights Laws in the Nineteen Western States*, vol. 1. p. 166.

20. *Bear River & Auburn Water & Min. Co.* v. *New York Min. Co.*, 8 Cal. 327, 332-333 (1857).

21. *Palmer* v. *Railroad Commission*, 167 Cal. 163, 170-171, 138 Pac. 997 (1914).

22. *Conger* v. *Weaver*, 6 Cal. 548, 556-558 (1856).
23. *Irwin* v. *Phillips*, 5 Cal. 140, 147 (1855).
24. *Hill* v. *Newman*, 5 Cal. 445, 446 (1855).
25. *Crandall* v. *Woods*, 8 Cal. 136, 144 (1857).
26. *Eddy* v. *Simpson*, 3 Cal. 249, 252, (1853).
27. *Tartar* v. *Spring Creek Water & Min. Co.*, 5 Cal. 395, 397-399 (1855).
28. Hutchins, *Water Rights Laws in the Nineteen Western States*, vol. 1. p. 170.
29. Johnson, *Remembered Yesterdays*, p. 82.
30. Wolfe, *The Life of John Muir*, pp. 270-275.
31. Johnson, *Remembered Yesterdays*, p. 288.
32. Ibid., p. 291.
33. Idaho: Idaho Laws 1881, p. 267. Kansas: Kans. Laws 1886, ch 115. Nebraska: Neb. Laws 1889, ch. 68; see Laws 1877, p. 168. North Dakota, Terr. Dak. Laws 1881, ch. 142. repealed N. Dak. Rev. Codes 1895, p. 1518, new enactment, Laws 1899, ch. 173. Oklahoma: Terr. Okla. Laws 1897, ch. XIX. Oregon: Oreg. Laws 1891, p. 52. South Dakota: Terr. Dak. Laws 1881, ch. 142. Texas: Tex. Gen. Laws 1889, ch. 88. Utah: Utah Laws 1880, ch XX; Laws 1897, p. 219. Washington: Wash. Sess. Laws 1889-1890, p. 706; Sess. Laws 1891, ch. CXLII. I am indebted to Hutchins for these citations.
34. Hutchins here cites *Forbes* v. *Gracey*, 94 U. S. 762, 766-767 (1877).
35. Hutchins here cites Wiel, S. C., "Water Rights in the Western States," 3d ed., vol. 1, 93 (1911).
36. Hutchins here cites *Broder* v. *Water Co.* 101 U. S. 274, 276 (1879).
47. Ibid.
38. Hutchins here cites *Jennison* v. *Kirk*, 98 U. S. 453, 459 (1879).
39. Hutchins here cites 16 Stat. 217 (1870).
40. Hutchins here cites 19 Stat. 377 (1877), 43 U. S. C. 321 *et seq.* (1964).
41. Hutchins here cites 26 Stat. 1096, 1097 (1891), 43 U. S. C. 321 et seq. (1964).
42. Hutchins here cites *California Oregon Power Co.* v. *Beaver Portland Cement Co.*, 295 U. S. 142, 160-163 (1935).
43. Hutchins, *Water Rights Laws in the Nineteen Western States*, vol. 1. pp. 171-175.
44. Ibid., p. 443.
45. *Jennison* v. *Kirk*, 98 U.S. 453, 459 (1879).

PART THREE

PRIVATE RIGHTS
IN FEDERAL LANDS

8

PRIOR RIGHTS AND THE FOREST

The stockman, who at times was the same person who had already acquired private property rights to the public land and waters through his activities in farming and mining, was following the same procedure to establish private property rights to the range that often surrounded his farming or mining property. The legitimacy of his actions was attested by the many efforts to spur Congress to pass a specific preemption law tailored to the greater acreage requirements of the western stockman. Mining interests had gained the 1872 Mining Act that had standardized the appropriation of mineral lands with the location-patent system of converting public lands to private property. The need for a similar law for stockmen had been recognized by the influential government official John Wesley Powell.* In 1878 he recommended—without success—that Congress pass the necessary laws standardizing settlement and appropriation of western grazing lands by stockmen.[1]

The battle for property rights continued on the grazing lands with stock water appropriation as the central factor in legal control of the range. As H. M. Taylor commented in the 1886 annual report of the Bureau of Animal Industry:

> It will be seen that the ownership of the watering places gives tenure to contiguous range. This fact is recognized by Western cattlemen, and the question as to the number of cattle individual owners are permitted to hold, under regulations of the various local associations, is determined by the questions of water frontage.[2]

The established stockmen continued to fend off the challenges of transient livestockmen and usurping capitalists who, in their mad scramble to lay claim to the resources of the western states, forced herds of livestock onto the ranges of the permanent settlers who had legitimate control of the water and therefore a legitimate claim to the surrounding grazing lands.

Abuse of the rangelands through overgrazing was predictable. Overgrazing was a direct result of northern core political efforts to obstruct the development of necessary laws to establish definitive private property rights on the vast acreages of open range: as we have seen, the cattle capitalists deliberately overstocked rangelands to drive off earlier settlers. Ironically, this overgrazing—which could only have happened in the absence of definitive private property rights—was later used by the conservationists and northern core financial interests as an argument in support of the nationalization of much of the West's land base through the establishment of forest reserves.

This nationalization continued through the years in several forms: first, the creation of national forests, then the Taylor Grazing Act of 1934, and most recently, the Federal Land Policy and Management Act of 1976. These acts respectively authorized the President to create forest reserves, extended federal

*Supra, p. 13.   149

grazing authority to non-Forest Service federal lands, and officially changed the policy of the United States to prevent the divestment of federal lands into private hands. Given the prior rights of users on many of these lands, it would be difficult to characterize this process as anything other than nationalization. It would be likewise difficult to show that this nationalization of the land base of the western states was not motivated by pressure to enhance the eastern industrial infrastructure. It was the result of failure to honor the pre-existing property rights that stockmen had established on the western range lands prior to 1891.

FENCES

Legitimate established stockmen for more than a hundred years have defended and strengthened their claims to preemptive property rights despite congressional inaction. Stockmen attempted to define their appropriative range rights and to protect their ranges from overgrazing by fencing their range.

Ominous signs of negative northern core attitudes toward the future of the West surfaced in the early 1880s as the federal government began a campaign to remove the boundary fences the established stockmen had erected. Complaints against fenced rangelands and petitions for relief piled into the General Land Office. The usual plea was from a newly arrived farmer claiming that the "evil cattle barons" were setting up "barbed-wire kingdoms to the great detriment of the intrepid pioneer." It did not matter to the newcomers that "pioneer" means the first one there. The General Land Office used the complaints as a basis for investigating unauthorized fencing. In 1882 and 1883 GLO Commissioner N. C. McFarland called for fence removal and the prosecution of ranchers by the Justice Department.[3] The GLO report was used to stir up Congress to pass an anti-fencing law, which it did in 1885.[4]

Strict enforcement of the law during the years 1885-1887 began with President Cleveland's order "that any and every unlawful enclosure of the public lands . . . be immediately removed."[5] Under the anti-fencing law the illegal fences began to come down. GLO Commissioner W. A. J. Sparks called on the president to use military force in removing fences.[6] In at least one place in Wyoming where a fence owner proved recalcitrant, federal troops were used to enforce the order.[7]

At the root of this fence removal effort was the naked objective of northeastern interests to dominate the West. They did not intend to honor the legitimate property rights of the early settlers.

Range rights by the late 1880s were well recognized as real property *according to local law and custom and court decisions.*[8] A western ranch unit in the arid states usually consisted of a small acreage in land that had been appropriated under the homestead or preemption laws and to which the owner had received a patent from the federal government, and vast acreages of rangeland on which a federal patent had not yet been issued. Control and legal ownership of the various springs, seeps, creeks, and wells on this federal land gave the rancher full control of the land. In the absence of clear congressional action to define property rights on the range, those rights had been thoroughly established according to local law and custom and court decisions. A common expression emerged in the West in reference to northeastern opposition to honoring definitive private property rights on the western range: "He who owns the water owns the land." The more arid the lands were, the more powerful were the implications of that statement. Westerners fully expected that the eastern-dominated Congress would be forced to acknowledge this reality.

Western ranches were bought, sold, traded, mortgaged, and inherited as a unit containing both fee simple land and the split estate federal lands. Local law and custom and court decisions fully accepted these transactions. Congress, while not acting to fully clarify the property rights issue on the western ranges, did acquiesce (as it had with the preemption of water), in the preeemption of farming lands and mining lands. The continued inaction of Congress was viewed by westerners as a clear signal that the stockmen's land claims would eventually be likewise honored.

Northeastern livestock interests had first tried to gain a prior appropriative claim to the West's resources by crowding out and disenfranchising the earliest settlers. They realized that control of stockwater on the federal lands meant control of the land itself. This preemption procedure could be expected to eventually lead to title to the land itself. By the late 1880s it had become apparent to the eastern capitalists that the early settlers still maintained a firm legal hold on the rangeland by ownership of water rights. This, coupled with the destruction of most of their livestock herds in the hard winter of 1886-87, caused the northern core interests to change course. If they could not gain ownership of the vast resource base of the western states, they would find a way to deny it to westerners. The forest reserve concept was seen as the tool by which the most desirable of the West's resource lands could be withheld from the citizens of the West.

MONOPOLY

The 1891 Forest Reserve Act created chaos and confusion throughout the West. As northeastern interests began to realize how powerful a tool had been placed in their hands by the addition of Section 24 and the repeal of the preemption laws in Section 2, the best of the West's resource base was rapidly locked away from westerners. As one reserve after another was proclaimed by Presidents Harrison and Cleveland, northeastern interests attacked the claims of stockmen in the most bitter manner. "Watershed protection" was used to take the best hydroelectric sites out of the hands of westerners, most of which sites were later developed by and for the benefit of northeastern interests. "Timber protection" fulfilled the purpose of preventing "cheap" competition for the northeastern timber barons. The best mineral lands were in the mountains of the West and in as much as Section 24 said that forest reserve lands did not have to be covered with forests, the best mineral lands were also in danger of being permanently locked away from westerners.

If the resources of the West could be permanently locked up until they were needed to support the established financial and industrial infrastructure of the Northeast, the objective of monopolizing the western states could be realized.

Was monopolizing, as such, actually in the minds of northern core politicians and the officials carrying out their policies on the western lands? We can gain a valuable insight from a proposal to Forest Service Chief Gifford Pinchot from a group of timber capitalists seeking to openly form a timber trust patterned after J. P. Morgan's billion dollar United States Steel Corporation. Pinchot responded by secretly offering to set up an invisible trust with the federal government as general partner.

The documentation for this astonishing move by Pinchot is preserved in a 1913 account by Philip P. Wells of the Forest Service Law Office, written at the request of Pinchot:

> After the panic of 1907 large operators in southern pine suffered from a severe depression. From their point of view the necessary thing was curtailment of production, but each one of them was afraid to reduce his output lest the others expand, or at least maintain theirs, and thus get all the benefit of his self-sacrifice. Therefore all the expensive mills were kept working full time, while stocks were accumulating and rotting in the lumber yards. Prices were too narrow to yield any profit except from logs that could be got out cheapest. This meant that the top log of each tree was left in the woods, stumps were cut high, fire risk was greatly increased, and there was much waste of timber in other ways.
>
> The lumbermen decided that they must form a trust after the fashion of United States Steel Corporation. They came to Washington to get assurance of immunity from prosecution under the Sherman Act and called upon President Roosevelt, conceiving their errand as one which chiefly concerned the Bureau of Corporations. But they had much to say of conservation as a reason and justification for their plans. Roosevelt referred them to you [Pinchot]. You told me of it one day after we had been at the White House, and we went from there to the New Willard Hotel for our (or, at least, my) first conference with the lumbermen.
>
> They were in great distress and wanted help but had no idea other than a lumber trust with "forest conservation" as sufficient excuse to guarantee their immunity. After leaving them I advised you that the power and duty of the Forest Service concerned *growing* timber, while lumbermen were only concerned in selling it; that they thought only of the market, while it was our business to think of the woods; that we should turn our backs upon and forget the market. Also we should make such arrangements with them as would insure conservative forest management of all their holdings: fire protection, careful logging, reproduction, and a gradual approximation to a sustained annual yield from the whole southern pine area so that new growth would, after a few years, balance the cut and a constant supply be produced from their holdings in perpetuity. Such a system of management would incidentally reduce the present cut somewhat and thus result in such a relief from overproduction as they were seeking. That would give them an inducement to submit to our regulations, but it was not our purpose in establishing or enforcing the regulation.

My method of bringing this about was not any trust, combination, or organization among them whatever, but a separate contract between the Forest Service and each one of them whereby each was to agree to cut and protect his holdings under our direction for the ends we had in view. Each was to contribute a certain sum per thousand board feet of timber cut (we at last settled on thirty-five cents) to pay the expense of our office and field force in administering the plan. This contribution would have yielded a greater sum each year than we were spending on the national forests.[9]

It is noteworthy that the timbermen came to President Theodore Roosevelt to obtain *immunity from prosecution* under the anti-trust laws. Roosevelt gave such immunity to other trusts. It is especially significant that these eastern timbermen cynically selected "conservation" as the justification for escaping prosecution. The "sustained yield" concept Wells mentions and that we today think of as a conservation measure in fact originated in just such calls to artificially restrict timber supply and bolster prices in a weak market. Even the sustained yield acts affecting the West resulted from timbermen lobbying Congress to restrict supply and support prices.[10]

One may wonder what difference there is between a trust in which the Forest Service is the controlling partner and a trust in which J. P. Morgan is the controlling partner, aside from the invisibility of Pinchot's proposed Forest Service trust. In any event, the lumbermen were too suspicious of Pinchot's motives to buy into his invisible trust. They went ahead with their own timber trust, "which Attorney General Herbert S. Hadley of Missouri (later governor) promptly 'busted' in the courts," wrote Wells, who may or may not have warned Hadley of the scheme.[11]

THE PREEMPTIVE RIGHT

There was one major barrier to the general forest reserve lockup effort: the stockmen's legitimate claim to range rights. A mining claim covered approximately twenty acres. A homestead usually covered only one hundred sixty acres. Range rights, however, covered almost every square foot of the West. There was virtually no part of the desert or mountains that had not been legitimately claimed by some stockman. Virtually all water on the federal lands had been claimed. To accomplish the complete lockup of the West's resources the ranchers' claims to range rights would have to be erased or controlled in favor of the reserve concept.

As we saw in earlier chapters, eastern interests first chose to ignore livestock grazing by the early settlers as a legitimate use of the unclaimed federal lands. They chose to portray the original settlers as trespassers and their claims to the lands and waters of the West as invalid. The U.S. Supreme Court opinion in *Buford* v. *Houtz* in 1890 changed all that.

The Supreme Court had clearly stated that permission had been granted by the federal government for the original settlers to use the range. One year after that decision a forest reserve clause was passed after almost twenty years of repeated defeats. The provisions of the 1891 Forest Reserve Act prevented any use whatsoever. As Gifford Pinchot himself noted with concern, the situation was "nearly impossible" in the reserves: "Legally, at least, no man could even set foot on a single foot of [reserve] land."[12]

CALLS TO REPEAL THE RESERVES

The attempt to ignore the legitimate range and water rights of established western stockmen nearly led to repeal of the forest reserves almost as soon as they were created. When Arkansas congressman Thomas C. McRae introduced a bill in 1893 to provide for some means of administering the forest reserves, he included a provision for the Secretary of the Interior to sell all merchantible timber on the reserves to the highest bidder. Westerners feared McRae's bill because it would allow a hostile Interior Secretary to deflect all timber away from poor settlers to rich combines and eastern capitalists. The westerners also realized that if they supported McRae's bill, it would be an implicit recognition of the power of Congress to legislate for the reserves, which would implicitly recognize the legitimacy of the reserves themselves. Westerners were not willing to take either risk.[13]

Colorado congressional delegate John Bell condemned McRae's bill and the forest reserve system in no uncertain terms:

> In the state of Colorado these reservations. . .take in farms, take in settlers, and in one county in particular they do not leave a single stick of lumber outside the reservation line, so from the day that reservation was declared every settler in the county has had to steal every stick that has gone into his fireplace. . . . Not only that, but there have been declared in a state reservation over a quarter million acres of land that has not a stick of timber or brush on it. . . . These reservations ought to be knocked out of existence. . . . In Colorado they have done us no good whatever.[14]

Angry westerners were not able to pass legislation repealing the reserves, but they did defeat all reserve legislation in 1894 and again in 1896. In light of *Buford* v. *Houtz* another U.S. Supreme Court decision on the legitimacy of range rights could well have sealed the doom of the forest reserves in their infancy. The political focus of the Bureau of Forestry and later the Forest Service would be to reach an accommodation with the stockmen short of formal recognition of private property rights in the range but sufficient to quiet the militant actions of the West's ranching community. Keeping the pre-existing range right controversy out of the federal courts was to become a cornerstone of federal land policy, as acknowledged by Will C. Barnes, one of the earliest employees of the Bureau of Forestry. He served as chief of the Grazing Service, U.S. Forest Service, from 1915 to 1928. In his *Reminiscences* he commented, "Great days were those for the Government Grazing policy; for say what you will, the grazing-men of the Forest Service were the shock troops who won the west for forestry."[15]

FIRST GRAZING PERMITS

The method by which this was accomplished can be summed up thus: the grazing permit system developed by the Bureau of Forestry was a tacit recognition of the pre-existing grazing and water rights of the permanent settlers, while official Forest Service policy was to deny that any pre-existing right existed.

A circular issued by the Department of Interior on January 8, 1902, stated that livestock on the forest reserves would receive preference in the following order:

1. Stock of residents within the reserve.
2. Stock of persons who own permanent stock ranches within the reserve but who reside outside of the reserve.
3. Stock of persons living in the immediate vicinity of the reserve, called neighboring stock.
4. Stock of outsiders who have some equitable claim.

The order of preference outlined in this U.S. Department of Interior circular clearly gave first preference to those ranchers who were residents and permanent. Those who were permanent residents in the livestock business in the West had become permanent by their prior appropriation of the water on the adjoining federal lands. Historian William Rowley commented about Albert F. Potter, first Grazing Chief of the Forest Service, illustrating the importance placed on private water in determining who received grazing permits.

> Probably because Potter was from the arid state of Arizona, he determined range capacity first by the proximity of water. The common method of range control throughout large areas of the West had been for local ranchers to secure title to land adjacent to the streams, springs, and lakes. By then denying the use of these sources to stray stock, stockmen often commanded thousands of acres of grazing land without purchasing or possessing title to it. Oftentimes, ranges with excellent forage possessed no watering places for stock. These situations called for such improvements as digging wells or building reservoirs. In some cases the water might be found in deep canyons that could be reached by building trails down the sides to the streams.[16]

About the time that Congress established the Forest Service in 1905 and gave it management authority over the forest reserves, Gifford Pinchot's staff published a "Use Book." On the subject of grazing the *Use Book* listed three classes of grazing permits:

A. For those who owned adjacent ranch property ("Small near-by owners.")
B. For those who owned nonadjacent ranch property and traditionally used the public forest ranges ("All other regular occupants of the reserve range").

C. for transient herders who could make no claim to local property ownership ("Owners of transient stock").[17]

The Forest Service adhered to the principle of "commensurate property ownership" as a condition for receiving forest grazing privileges. The concept of commensurability suggested the ownership by the permittee of enough private ranch land to support the stock during periods when for one reason or another the national forest was not open to grazing.[18]

A and B permits as noted in the "Use Book" were commensurable (co = with + mensurable = measurable; *i.e.*, "measurable with") with private fee-simple land. The private "commensurable" land was that of permanent residents. Those ranchers became permanent residents in their respective areas by virtue of the fact they had a legal pre-existing right to the stock water on the surrounding federal range lands. Most of those commensurable pieces of private land were totally unsuitable for anything but supporting livestock on the range. Topography, elevation, and climate insured that. The difference between a permanent resident and transient user of the range was largely determined by who could show prior appropriation to the stock water on the range. The proof of beneficial use for the springs, seeps, and creeks on the federal lands was livestock grazing. Livestock grazing was a legitimate right that existed prior to the forest reserves as attested to in *Buford* v. *Houtz*. The Forest Service permit system then became an exercise in granting permits to those who could prove they had a pre-existing right to graze while at the same time attempting to deny that any pre-existing right existed.

TWO VIEWS OF "CONSERVATION"

The year 1905 was a watershed in the development of the national forest system. Congress created the Forest Service in the Department of Agriculture and transferred the forest reserves to its authority. President Theodore Roosevelt appointed Gifford Pinchot as first Chief of the Forest Service. Pinchot and his utilitarian vision of conservation now had real power. By extension the northern core political economic interests had accomplished a major step in permanent monopolization of the American West.

Pinchot had already broken with preservationists such as John Muir and was prepared to push the utilitarian vision of conservation to the exclusion of preservation if necessary, as reflected in the 1908 battle over Hetch-Hetchy reservoir near Yosemite. Pinchot sided with the pro-reservoir forces who wanted the Sierra Nevada valley flooded to provide domestic water for San Francisco.

The break with Muir had come as the result of a chance meeting in a Seattle hotel lobby in 1897. It was the result of a violent disagreement over sheep grazing: Muir had discussed the undesirability of sheep on the forest reserves with Pinchot on an earlier field trip and thought that the forester had agreed that sheep should be forbidden on the reserves. Then Pinchot told a newspaper that sheep weren't so bad after all. Muir used the chance meeting to tell off Pinchot in florid language. Muir biographer Linnie Marsh Wolfe describes the acrimony:

Stopping in a Seattle hotel on the way home, Muir read in a morning paper that Gifford Pinchot, then in the city, had given out a statement that sheep grazing in the reserves did little if any harm. Now, it happened that the most potent politicians of the state were either sheep and cattle men, or closely allied with live-stock corporations. So along came Mr. Pinchot with his little appeasement policy!

This struck Muir like a blow in the face. Not only did it shake his faith in a man he had trusted, but to his mind it augured ill for conservation, since Pinchot, recently appointed by Cornelius Bliss as Special Forestry agent, was here to make a survey of the commercial resource of the suspended Cleveland reserves.

During the morning Muir saw Pinchot in the hotel lobby, standing a bit apart from a group of men who—unbeknown to Muir—were newspaper reporters. According to William E. Colby, to whom he related the incident, Muir went up to Pinchot and, thrusting the printed page before him, demanded:

"Are you correctly quoted here?"

Pinchot, caught between the two fires, had to admit that he was.

"Then," said Muir, his eyes flashing blue flames, "if that is the case, I don't want anything more to do with you. When we were in the Cascades last summer, you yourself stated that the sheep did a great deal of harm."

Thus the rift opened that swiftly widened between the two schools of conservationists—the strictly utilitarian, commercial group who followed Pinchot, and the aesthetic-utilitarian group who followed Muir—a rift that was to manifest itself deplorably in long years of antagonism between two Government bureaus.[19]

The northern core interests with whom Pinchot was associated had been forced to permit grazing in the forest reserves to avoid potential loss of the reserve system and the full recognition of the rancher's pre-existing property right. To John Muir, who had been used by the same northern core interests for years to preach against commercial use of the forests—livestock grazing in particular—Pinchot's action was a betrayal.

Muir's personal eccentricities had become pronounced over the years. Linnie Marsh Wolfe discusses Muir's idiosyncrasies in her book *Son of the Wilderness*:

But as if for his benefit, Nature herself staged two stupendous spectacles during those months. The first was a storm in mid-December. For two days and nights "the jubilee of waters" lasted. He was abroad every hour in the thick of it, reveling in the wild elemental mingling of winds, waters, and rocks. . . .

Nature's second grand show was an earthquake.[20] Sleeping in his cabin on the night of March 26, he was awakened by "a strange wild thrilling motion and rumbling." Disregarding the hard fist of fear in the pit of his stomach, he ran out into the moonlit meadows. But for once he couldn't run far. The earth twisted and jerked under his feet. There was an infernal rumbling like the pounding of giant hammers from beneath.

Then after a moment of calm came the climax. Eagle Rock, high on the south wall of the valley, was toppling, then falling like a fiery comet out of the sky. All fear forgotten, he bounded toward the descending mass, and before the huge heated rocks had settled, groaning, into their places on the valley floor, he leaped upon them in a shower of dust and falling fragments, shouting: "A noble earthquake, a noble earthquake!" . . .[21]

Muir himself bragged about his fearlessness in the face of nature's danger, as Wolfe summarized from one of his books:

Leaving Shasta at the end of December, Muir went to Brownsville. . . The morning after his arrival he donned his host's old coat and pushed out into the woods to see how the trees behaved in a wild gale. The storm was holding high carnival when he climbed to the top of a Douglas spruce towering above the surrounding forest. The tree beneath him rocked and swirled, forward, backward, around and around. No bucking mustang ever gave him so mad a ride. But with iron spikes in his shoes and steel muscles in hands and arms, he clung to his wild steed as lightly as "a bobolink on a reed."[22]

Muir's biographer also emphasizes his religious beliefs.

Muir was deeply impressed with the fact that to the Indian mind all Nature was "instinct with deity."' And although deity was personalized in spirits that inhabited mountains, rivers, and waterfall, he with his own deep-seated paganism, felt these children of the wilderness came nearer to the truth of an immanent living Principle in all matter, than did the tutored, civilized exponents of Christianity. [23]

Muir's brilliance and charisma was noted early on by the proponents of national parks. Robert Underwood Johnson, editor of *Century* magazine, encouraged Muir's writings in support of land withdrawals from the public domain for conservation purposes. Johnson, born in Washington D.C. and educated in Indiana, was instrumental in introducing Muir to the northern core political financial interests.

Muir's genius on the one hand was apparently offset by a vulnerability to the very wealth and materialism he continually decried. As long as Muir was needed for the purpose of attacking the stockmen, he was courted by the northern core elite in grand fashion.

Wolfe describes Muir's introduction to the eastern movers and shakers while on the first leg of a world tour. Muir traveled with his close friend William Keith, an artist from San Francisco:

Muir hastened on to New York, only to find his fellow Scot being whirled from one reception to another, staged for him by the elite of society and the art world.

Muir settled down in an obscure hotel to wait for Keith and finish some articles he had brought along. But Robert Underwood Johnson had other plans. Whisking him bag and baggage over to the Players' Club, he began to introduce him to celebrities. There ensued such a succession of luncheon and champagne dinners given in his honor, "I had no idea I was so well known, considering how little I have written," he naively told the home folks.

On a blazingly hot day Johnson took him for a cool trip up the Hudson to Castle Rock, the beautiful country home of Henry Fairfield Osborn, the paleontologist. This began one of the closest friendships of his life, with the Osborn family.

Then came visits to Boston, Cambridge, and Concord. In the last-named place he and Johnson tramped about among the homes, haunts, and graves of Emerson and Thoreau. They walked to Walden Pond, where "the blessed crank and tramp" had kept his rendezvous with wild nature and worked out his own "experiment."

On another day they went to Brooklyn to visit Charles S. Sargent, the Harvard botanist. Here Muir was entertained at dinners formal and informal. . . .

He returned to New York "crowned and overladen with enjoyments," to be dined and wined by Dana of the New York *Sun*. At other affairs he met Mark Twain, Thomas Bailey Aldrich, Charles Dudley Warner, Rudyard Kipling, George W. Cable, and Nicola Tesla. On one occasion he was feted "in grand style" by Mr. James Pinchot of Gramercy Park, "whose son is studying forestry."[24]

E. H. Harriman of New York—and of the Union Pacific and Southern Pacific Railroad—was one of the leading capitalists of his day. He provided Muir with extensive prepaid scientific expeditions, free rail fare throughout the nation, and in 1902, a free round-the-world trip. Harriman and his northern core political interests kept Muir in their service by generous red carpet treatment. Muir's obvious and sometimes embarrassing eccentricities were explained away as the mark of genius.

After the reserves had passed the repeal challenge in the 1893-1902 period, Muir and the Sierra Club he co-founded were of decreasing importance. Once the utilitarian arm of the conservation movement gained control and the danger of forest reserve repeal had largely subsided, Pinchot could begin pushing Muir and the preservationists out of the movement's mainstream. Pinchot says of the nineteenth century conservation impulse, "the handful of forward looking men and women" worked with genuine enthusiasm but "unfortunately with very little result. They spread," he said, "a doctrine which they called Forestry but which in fact was forest preservation—a very different thing."[25]

Pinchot had tremendously good luck with the forest reserves. A good part of that luck came from what is known today as the Sargent Commission. Headed by eminent Harvard dendrologist Charles S. Sargent, the commission was convened in February 1896 at the request of Interior Secretary Hoke Smith to the National Academy of Sciences for a study on whether fire protection and permanent forests were practical on the public domain; to estimate the influence of forests on climate, soil, and water; and recommendations for specific legislation. Pinchot and Robert Underwood Johnson convinced a meeting of the American Forestry Association—over the objections of Bernhard Fernow—to push the government for such a commission.*

The members of the commission included Professor Alexander Agassiz of Harvard—a large investor in N. R. Davis Cattle Company in Wyoming and Nebraska, Union Cattle Company of Cheyenne, the Union Pacific mines at Telluride, Colorado, farms in Colorado and real estate in California—Army engineer Henry L. Abbott, William H. Brewer of Yale, Arnold Hague—who convinced Interior Secretary Noble to recommend the first forest reserve to President Harrison—and Gifford Pinchot, the only one who was not a member of the National Academy of Sciences. Muir joined the party in the West as an unofficial participant.

The commission traveled to the West, made their investigations, and recommended the immediate creation of thirteen reserves covering 21 million acres. Forest reserves already covered 20 million acres of the West. The commission suggested that Washington's birthday, 1897, would be an appropriate date for the president's proclamation. President Cleveland obliged.[26]

The "Washington's birthday reserves" set off a furor in the West that refused to abate for more than a decade. Pinchot spent much of his federal career trying to save the reserves from repeal because of that sudden and vast set-aside.

The success of Pinchot in defusing the potentially disastrous opposition of the western ranchers was largely the result of his granting of permits to graze based on pre-existing rights and by employing westerners to oversee the new grazing program on the reserves. Albert F. Potter, an Arizona sheepman, was appointed grazing expert in the Bureau of Forestry in 1901. After the Forest Service was created in 1905, he served as Grazing Chief until 1910. Others included were: W. C. Close who had spent much time in Utah studying and developing grazing procedures; C. H. Adams was formerly in the livestock business; Joe Campbell, formerly had been a livestock inspector in Arizona; Leon F. Kneipp of Arizona; Will C. Barnes, once manager of the Esperanza Cattle Company.

By chosing westerners to develop and implement the rules governing grazing on the national forests, the objections to those "scientific men from Harvard College"—the Sargent Commission—was blunted. Westerners, many of them stockmen themselves, understood and could operate within the pre-existing rights of the ranchers on the reserves with minimum conflict. An outsider, such as the "scientific man from Harvard College," may inadvertently refuel the preemptive property rights argument of the ranchers and bring on the much feared court test of that doctrine. Pinchot and the northern core interests

*Cf. supra, p. 118.

were doing in the colonial West what any wise occupation force has historically done—enlist locals to carry out the enforcement of foreign edicts.

UNEASY ALLIANCE

On December 12, 1905 a joint meeting of representatives of the new Forest Service and stockmen was held in Denver at the Albany Hotel. The stockmen who owned pre-existing rights in the forest reserves were organized as the American National Cattlemans Association. Their purpose was to form a united front to defend their property rights on the nationalized forest lands in their respective states. The new Forest Service was there to discuss the joint management of grazing in the national forests. The stockmen were there to tie down permanent grazing rights in the national forests. The Forest Service was there to discuss how they were going to recognize the grazing rights of the established ranchers according to priority of use without acknowledging the existence of preemptive private property rights.[27]

At this meeting Pinchot and the new Forest Service were dealing almost entirely with the original settlers, their heirs and assigns. The northeastern cattle capitalists who had originally favored a reserve system as a way to monopolize the West's best grazing lands had almost been eliminated from the picture by the harsh winters of the late 1880s. The survivors tended to be those ranchers who owned prior rights. It was these early established settlers who had been best prepared to deal with the vagaries of nature.

The prior rights issue addressed at the Denver meeting had been at the root of the 1897 protest by western congressmen. That protest had resulted in a rider being attached to the general appropriations bill demanding the repeal of the reserves. President Cleveland refused to sign the bill. Subsequently the federal government was forced to operate without funds until President McKinley took office in March of that year. McKinley took immediate action to accommodate western interests so the appropriations bill could be passed.[28] That accommodation opened the door for the official acceptance of livestock grazing as a legitimate use of the forest reserves.

At the 1905 Denver meeting,

> The main points of agreement, worked out by the department and stock organizations, emphasized that those already grazing in the forest ranges would be protected in their priority of use; that reductions in the number of grazed stock would be imposed only after fair notice; that small owners would have preference over large; that only in rare circumstances would the department seek total exclusion of stock from the forest; and that the policy of use would be maintained wherever it was consistent with intelligent forest management. Finally, some attempt would be made to give stockmen a voice in making the rules and regulations for the management of stock on local ranges through the establishment of forest advisory boards.[29]

It was this 1905 meeting where the alliance between the Forest Service and the livestock associations first emerged. The alliance continues to this writing in 1989. This alliance between the Forest Service and the stockmen would be maintained by Forest Service domination of the leadership of the livestock associations.   Concessions to influential stockmen in the national forests prompted Gifford Pinchot to comment in 1907 that, "The Forest Service grazing program was developing 'a constituent group' that found security and profit in the government program."[30] Lawrence Rakestraw explained in his *History of Forest Conservation in the Pacific Northwest 1891-1913*, that livestock associations often served the function of "company unions" for the Forest Service by supporting its policies and complying with regulations in return for concessions allowing certain stockmen to purchase and absorb smaller allotments or needed range improvements.[31]

Open rebellion was often simmering just below the surface of the alliance. The resentment of the stockmen toward the federal government for failure to openly honor their preemptive property rights led to numerous conflicts. Stockmen continued to buy, sell, trade, inherit, and mortgage range rights as part and parcel of the ranch unit. Banks and lending institutions continued to view range rights as real property and collateralized those range rights for loan purposes. The grazing rights on the split estate lands of the national forests obviously enhanced the value of the ranch.[32]

The Forest Service feared the possibility that a stockmen might be able to sue in defense of his rights and thereby establish in the courts that a preemptive private property right did in fact exist.

The bottom line as far as the Forest Service was concerned was that recognition of the rancher's preemptive property rights and survival of the U.S. Forest Service were mutually exclusive. The preemptive rights argument had to be contained at any cost.

To contain the property rights argument the Forest Service began to curtail private investment in range improvements by the ranchers. Private investment in the fencing, water facilities, livestock handling facilities, roads, or any other range improvements increased the rancher's private property right claims. The Forest Service countered this threat by making the improvements for the rancher with public (taxpayer's) money thereby increasing the "public" interest in the split estate lands. Many preemptive property rights challenges from individual stockmen have been bought off over the years with public monies in the form of range improvements.

The persistent assertion on the part of the ranchers that they owned their permit was viewed with alarm. To counter the threat to bureaucrats represented by widespread recognition of the underlying property right held by the stockmen on the national forests, the Forest Service began to refer to grazing permits as only a "privilege" obtained from the Secretary of Agriculture.[33]

Such verbal obfuscation used in an attempt to prevent general public recognition of the stockmen's property rights infuriated the ranchers while at the same time doing nothing to prevent the treatment of grazing rights as real property in ranch business transactions. The livestock associations, playing the role of company unions, were very instrumental in quieting these conflicts. Cooperative ranchers could gain obvious financial advantage from the forest grazing programs. Uncooperative ranchers could find themselves facing the joint pressure of fellow stockmen, the stockmens' associations, and the Forest Service. After one particularly wrenching confrontation with the stockmens' rights issue in Montana in 1916—that had been quieted through the cooperation of the Forest Service and the livestock associations—a Forest Service official wrote, "the radical feeling had died down and we seem to be on good terms with our worst enemies."[34]

As a further attempt to discourage recognition of the rancher's pre-existing rights, the Forest Service established the permit waiver system. Theoretically the waiver of the permit back to the Forest Service upon sale or transfer of a ranch divested the rancher of any acquired property rights in his permit. The Forest Service then reissued the permit to the new owner of the ranch as a privilege granted by the federal government. This procedure has been used as the basis of an argument to assert total Forest Service ownership of range rights.

The flaw in this argument is that the permit had originally been granted in recognition of real property rights appropriated by the rancher in the range before March 3, 1891. The permit was a recognition of real value owned by the rancher, the value of the pre-existing range rights and water rights. Permits based on a certain number of animal-unit-months (AUM's) set a value on those pre-existing rights. Simply put, an animal unit month is the amount of feed required by a cow for one month.[35] It was this AUM value that was basis for determining the grazing fee.

The waiving of the permit back to the Forest Service was in fact only waiving the current value the Forest Service had placed on the pre-existing rights. The permit waiver was an acknowledgement on the part of the seller that the Forest Service had the right to renegotiate the number of AUM's the new owner would be responsible for on his particular forest allotment. The permit waiver gave the Forest Service the privilege of renegotiating the current value of the underlying right for the purpose of taxation.

I use the term taxation in discussing the grazing fee for two reasons:

1. A fee is a form of a tax, tax being the broad term and fee being the narrow term. "Fee" is usually used in restrictive circumstances such as a less-than-year-long grazing season.

If the use is continual and unbroken the term tax is usually applied.

2. When the grazing fee was first instituted on the forest reserves in 1906, it was often referred to as a tax.[36]

Certainly the western rancher, who had spent years appropriating land and water from the public domain and whose primary objective prior to 1891 was to establish private property rights in the federal lands, was thinking in terms of taxes when grazing charges were first discussed. The rancher, who had been granted a permit to graze in the new forest reserves, had received that permit on the basis of proving he had a prior right to the water and grazing in the reserve. Claimants unable to prove prior use or ownership of commensurable land and water did not receive permits. The rancher being charged for grazing under the permit system viewed it as a property tax paid to the federal government. He paid property taxes on his fee-simple property (his commensurable base property) to the local government. The grazing charges were instituted to pay the cost of government administration of the split estate land. Taxes paid to local government served essentially the same function in relation to fee simple property.

LESSONS LEARNED

One of the major weaknesses in the stockman's prior right argument was a lack of statutory support for his claims. The prior appropriation-vested rights argument was often supported only by the testimony of living witnesses. While some water claims had been properly recorded with local authorities, many claims stood or fell on the availability of living witnesses who could verify that the claimant or his predecessor(s) was in fact the first to use a certain water on the federal lands for stock water and therefore put the water and range to beneficial use.

This was a part of "local law and custom and court decisions" which both the U.S. Congress and the U.S. Supreme Court had upheld. In 1891 it was still fairly easy to find three people who had been in the area for twenty or thirty years and could swear to first use appropriations.

The first major land withdrawals in the West were the federal grants to California for a state park in the Yosemite area in 1864 and the creation of Yellowstone Park in Wyoming in 1872. Development of state statutes to protect appropriative rights on the federal lands began to accelerate with the conservationist clamor for forest reserves and more parks during the 1880s. With the federal plunder of the West's richest resource lands starting in 1891, there was an even greater effort on the part of the states to develop statutes governing non-navigable waters within their borders.[37]

Passage of the 1891 Forest Reserve Act prompted states to formalize their statutory procedures for recording water rights and assigning private title to those rights. Every proposal to create a new forest reserve or national forest prompted stockmen to avail themselves of the new formalized procedures. Particularly after 1890, state records show a significant increase in formal water filings by ranchers on their appropriative claims with every threat of a new forest reserve or national forest. Stockmen were racing to tie down their one indisputable private property right on the new forest reserves and national forests: their stock water rights.

Chapter Eight Footnotes

1. Powell, *Report on the Lands of the Arid Region of the United States*, pp. 21-29.
2. H. M. Taylor, "Importance of the Range Cattle Industry," *Annual Report* of the Bureau of Animal Husbandry, Washington, 1886, p. 316.
3. Report of the Commissioner of the General Land Office on the Unauthorized Fencing of the Public Lands, *Sen. Ex. Doc.* No. 127, 48 Cong., Sess. 1, 1883-1884.
4. 23 Stat. 321.
5. Gates and Swenson, *History of Public Land Law Development*, p. 468.
6. William Sparks, General Land Office Commissioner, to Interior Secretary Lucius Lamar, August 19, 1885. R. G. 48, Office of the Secretary of the Interior, Central Classified Files, National Archives.
7. Gates and Swenson, *History of Public Land Law Development*, p. 474.
8. R. Taylor Dennen, "Cattlemen's Associations and Property Rights in Land in the American West," *Explorations in Economic History*, vol. 13, pp. 423-436.
9. *Forest History*, "Philip P. Wells in the Forest Service Law Office," April, 1972, pp. 26-27.
10. David T. Mason, "Sustained Yield and American Forest Problems," *Journal of Forestry*, vol. XXV, No. 6, October, 1927.
11. Ibid., p. 27.
12. Pinchot, *Breaking New Ground*, pp. 85-86.
13. McCarthy, *Hour of Trial: The Conservation Conflict in Colorado and the West, 1891-1907*, pp. 48-49.
14. *Congressional Record*, 53, Cong., 1 Sess., vol. XXV, Pt. 2, pp. 2434-35.
15. Will Croft Barnes, *Apaches and Longhorns: The Reminiscences of Will C. Barnes*, University of Arizona Press, Tucson, 1982, p. 202.
16. Rowley, *U.S. Forest Service Grazing and Rangelands: A History*, p. 69.
17. U.S. Forest Service, *The Use of the National Forest Reserves*, July, 1905, p. 22.
18. For a discussion of "commensurate property" see Rowley, *U.S. Forest Service Grazing and Rangelands: A History*, p. 69.
19. Wolfe, *Son of the Wilderness: The Life of John Muir*, pp. 275-276.
20. Wolfe here cites "Yosemite in Spring" (letter of May 7, 1972), New York *Tribune* (undated clipping).
21. Ibid., p. 157.
22. For Muir's forest storm experience see "A Windstorm in the Forests," *Mountains of California*, vol. I, Houghton Mifflin, Boston, 1917, pp. 272-86.
23. Based upon Muir's fragmentary notes as represented by Wolfe in *Son of the Wilderness*, p. 209.
24. Wolfe, *Son of the Wilderness: The Life of John Muir*, p. 262.
25. Pinchot, *Breaking New Ground*, p. 92.
26. For a thorough telling of this story see Steen, *The U.S. Forest Service, A History*, pp. 30-34.
27. Paul Henle Roberts, *Hoof Prints on Forest Ranges: The Early Years of National Forest Range Administration*, Naylor, San Antonio, 1963. p. viii.
28. Rowley, *U.S. Forest Service Grazing and Rangelands: A History*, p. 30.
29. Albert F. Potter, "Cooperation in Range Management," *American National Cattleman's Association Proceedings*, 16 (1913), p. 55.
30. Rowley, *U.S. Forest Service Grazing and Rangelands: A History*, p. 74.
31. Lawrence Rakestraw, *History of Forest Conservation in the Pacific Northwest 1891-1913*, New York, 1979, p. 81.
32. Rowley, *U.S. Forest Service Grazing and Rangelands: A History*, pp. 89-90.

33. See *Annual Grazing Report, 1916*, Rio Grande National Forest, Sec. 63, Region 2, Dr. 35, RG 95, National Archives.

34. *Annual Grazing Report, 1912*, Sopris National Forest, Sec. 63, Region 2, Dr. 31, RG 95, National Archives.

35. For a technical discussion of AUMs see Rowley, *U.S. Forest Service Grazing and Rangelands: A History*, pp. 164-165.

36. The notion of grazing taxes is discussed at length in McCarthy, *Hour of Trial: The Conservation Conflict in Colorado and the West, 1891-1907*, pp. 161-164. The term "tax" was used almost universally in newspaper reports in the West from 1905 for a decade or more.

37. Hutchins, *Water Rights Laws in the Nineteen Western States*, vol. 1, pp. 170-71.

9

PRIOR RIGHTS AND THE COURTS

The case of *Kansas* v. *Colorado,* decided by the U.S. Supreme Court in 1907, had a significant impact on the attitudes of federal agents in the newly created Forest Service. The case itself was a dispute between Kansas and Colorado over who had jurisdiction over the water of the Arkansas River which originated in Colorado, an appropriation doctrine state, but crossed Kansas,a riparian doctrine state. The future of the West's water rights doctrine lay in the balance.

Colorado and the appropriation doctrine won the case, which had raised the question of states' rights and the ability of the states to control the resources within their own borders. The states were particularly sensitive to federal intrusion on their rights given the massive forest reserve withdrawals since 1891. The Forest Service treated the reserves as federal property subject only to jurisdiction of the federal government. The use of federal troops to enforce national park policy in Yellowstone, Sequoia and General Grant National Park further aggravated the strained relationship of the states and federal government. The federal authorities in charge of the national forests asserted the doctrine of sovereign immunity. They viewed the withdrawn forest lands as lands subject to the provisions of Article I, Section 8, Clause 17 of the U.S. Constitution:

> To exercise exclusive legislation in all cases whatsoever, over such district (not exceeding ten miles square) as may, by cession of particular states, and the acceptance of Congress, become the seat of the government of the United States, and to exercise like authority over all places purchased by the consent of the legislature of the state in which the same shall be for the erection of forts, magazines, arsenals, dockyards, and other needful buildings.[1]

This clause provided for state approval of land withdrawals by the federal government within state borders. Forest reserves were proclaimed without state approval and usually over strenuous protest from the states.

From the beginning though, the forest reserves had been administered by the federal government in general disregard of the laws and authority of the states. This was, of course, in line with the political objective of forest withdrawals. If the choicest hydroelectric sites were to be taken out of the hands of westerners for the benefit of northeastern industry and capital under the guise of watershed protection, western water rights had to be extinguished or neutralized. If the development of irrigated agriculture in the arid West were to be controlled by northern core interests, control of the water was a must. If northeastern industrialists and financiers controlled the water needed for the development of western mines, their control of the output of the mines would increase. The processing of trees into salable lumber from the western forests also depended largely on water. Stockmen needed water to raise livestock. If control of water on the forest reserves could be obtained, the stockmens' preemptive range rights, which blanketed almost every square mile of the reserves, could be eliminated.

Wresting control of the water from the states was a primary objective of national forest administrators. If the federal government had legal sovereignty over forest reserves as federal enclosures, western water law would be gutted. Virtually all the water in question originated in areas withdrawn under Section 24 of the 1891 Forest Reserve Act. If the forest reserves were legally sovereign federal enclaves, then the total subjection of the West would be complete. The northern core would not have to fear the loss of political and industrial dominance posed by a rapidly developing, resource-rich West.

The decision in *Kansas* v. *Colorado* dealt a severe blow to the ambitions of northern core proponents of western state subjugation. Colorado's victory also clarified the status of the forest reserves and their waters within the states.

The *Pueblo Chieftain* on May 15, 1907, reported the court's decision:

> That the land belonging to the national government within the states is subject to the law of those states.
>
> That the only control of the waters of an interstate stream by congress is in respect to the question of navigation.
>
> That a state has full jurisdiction over all lands and non-navigable streams within its borders and that it may determine which law shall govern with respect to such lands and waters, whether the common law rule of riparian rights or the doctrine of appropriation of water. . . .[2]

The *Chieftain* then editorialized:

> The principle of public ownership and public control of water, subject to appropriations for beneficial use according to priority, embodied in the constitution of Colorado and other western states, is recognized and approved by federal supreme court.
>
> The attempted usurpation of federal power by congress and the executive departments growing out of the federal possession of public lands, is properly rebuked and nullified.[3]

FOREST SERVICE REACTION

Reaction of the Forest Service and its supporters to the *Kansas* v. *Colorado* decision was predictable.

General William J. Palmer, who represented northern core interests in western railroad ventures, responded with considerable alarm to the decision in a letter to Gifford Pinchot on May 23, 1907:

> Dear Mr. Pinchot,
>
> I enclose an article from a Colorado paper [the *Chieftain*] purporting to give the substance of a recent decision of the Supreme Court of the United States, which is being much harped on by the newspapers generally in this part of the continent. I can scarcely believe that the Court intended its

decision to be so far reaching, but I think even if they did not, the dissemination of such views may have a very deleterious effect on the Forestry policy of the United States. . . .[4]

General Palmer also wrote to Philip P. Wells, law officer for the Forest Service, Department of Agriculture, Washington D. C., urging the Forest Service to resist the Supreme Court decision.*

> I trust that the gentlemen representing the Government Forestry Department who may attend the Convention at Denver, will yield no point whatever that the law, as interpreted by the Courts, has not already determined, but insist on the same rights over wood, grass and water on its own reservations, that the State of Colorado does. . . . The battle may as well be fought to the finish now as later.[5]

Wells alerted Thomas E. Will, Secretary of the American Forestry Association in Washington D.C. The American Forestry Association was one of the conservation groups whose efforts had been instrumental in the creation of the forest reserves. The AFA periodical *Forestry and Irrigation* conveyed the views of northern core interests to the public. *Forestry and Irrigation* was a widely quoted source for the general media on the forest reserve issue. Its effectiveness in creating the reserves and helping stem the repeal effort had been significant. A letter to Dr. Will written by acting law officer of the Forest Service, G. S. Arnold, on June 24, 1907, contained the following comment:

> I am under the impression that the interpretation of the case . . . is . . . rather too broad, so far as it admits restrictions upon the *absolute government control of the water upon Government land.*[6] [My italics.]

The U.S. Supreme Court had effectively unravelled the northern core's efforts to gut western water law. Wells wrote to Dr. Will on June 29, 1907 expressing alarm over the water issue and his reluctance to accept the U. S. Supreme Court decision upholding the prior appropriation water doctrine of western states.

> The critics of the Forest Service and the Reclamation Service, without any justification as it seems to me, construed this decision to mean that Congress has no legislative power over the land of the United States within a State for the purpose of regulating water rights upon, and the reclamation of such public lands.[7]

The "public relations arm" of the Forest Service responded to the alarm. In a letter to G. S. Arnold, the American Forestry Association's Dr. Will wrote:

My Dear Sir:
I have your letter of June 24th together with copy of Mr. Wells' letter

*Cf. supra, pp. 152-53 for Wells' role in the Forest Service.

to General Palmer. I note, that on Mr. Wells' return, he may desire to give the matter further consideration. If, when this has been done, I could have a statement from your office, either the Palmer letter or such modification as Mr. Wells may think proper, I should be pleased to give the matter some space in "Forestry and Irrigation."

Thanking you for your help in this matter, I remain,
Very truly yours,
Thomas E. Will
Secretary[8]

With the case of *Kansas* v. *Colorado* an unwritten policy developed within the Forest Service regarding state water control and private stock water rights in particular. It grudgingly complied with state water law while avoiding formal recognition of it, all the while devising methods to undermine or weaken it.

An extension of this philosophy is expressed in Forest Service treatment of the terms *prescriptive* and *preemptive* when discussing the range rights claims of ranchers. Open acceptance of court rulings on the states' control of stock water within the national forests and the ruling in *Buford* v. *Houtz* that grazers had an implied license to use federal lands would demand use of the term *preemptive* when discussing the rancher's claim to grazing rights. Preemption implies that the rancher's claim results from federal recognition of prior rights. This recognition, if accepted by the Forest Service, would argue against the very existence of the forest reserves and national forests.

This they refuse to do. Chief Forester William B. Greeley in 1925 pointed out that the Department of Agriculture had always recognized the rights of prior users of grazing lands. "It has never, however," he declared, "been able to sanction the claim to a vested right of property interest in the forage. On the contrary, it has regarded that conception as absolutely inimical to the principles of conservation."[9]

This contradictory position on the rancher's range rights is expressed in the use of the terms *prescription* or *prescriptive*. Prescriptive rights are rights by use without permission of the sovereign. Forest Service adherence to the term *prescription* is a manifestation of the original philosophy behind the first forest reserves. It is a denial of the prior appropriation water doctrine and the holding of the U.S. Supreme Court that grazers have an implied license to use the federal lands.

RESURGENCE OF PREEMPTIVE RIGHTS

Part of the concern expressed by the Forest Service over the *Kansas* v. *Colorado* decision stemmed from what it feared could be a threat to their newly acquired control over livestock grazing. The carefully worked-out compromise between the Forest Service and stockmen who held pre-existing rights on the national forests hung in the balance. The U.S. Supreme Court ruling had been clear, and as much as the proponents of the national forests hated to admit it, the states had sovereignty over the national forests. State water law would have to be observed.

The implication of this, as far as grazing on the national forests was concerned, was a strong reinforcement of the stockman's pre-existing rights.

The stockmen's preemptive private property rights again threatened the entire national forest system. If stock water rights created by the states on federal land—using prior grazing use as proof of beneficial use—were valid, then the U.S. Supreme Court was very close to recognizing the private property rights in the range which the stockmen and state governments had been attempting to create prior to March 3, 1891.

The Forest Service and its supporters worked diligently during this period to quiet the preemptive property rights challenge by winning over stockmen to their side. This was done by giving the stockmen who held valid prior-right claims a privileged position in securing and expanding grazing permits. The stockmen who could prove they had a pre-existing right to graze in national forests were given a preference over other applicants for grazing permits. If this favored treatment did not make national forest proponents out of the privileged stockmen, it for the most part quieted dangerous opposition.

Another effective tool in containing the preemptive rights issue was the creation of advisory boards. These advisory boards were often the local livestock association. Regulation 45 in the 1910 "Use Book" read as follows:

> Reg. 45 Whenever any live-stock association whose membership includes a majority of the owners of any class of live stock using a National Forest or portion thereof shall select a committee, an agreement on the part of which shall be binding upon the association, such committee upon application to the district Forester, may be recognized as an advisory board for the association, and shall then be entitled to receive notice of proposed action and have an opportunity to be heard by the local Forest officer in reference to increase or decrease in the number of stock to be allowed for any year, the division of the range between different classes of stock or their owners, or the adoption of special rules to meet local conditions.[10]

By allowing a select group of stockmen—those with preemptive private rights claims—to essentially control the grazing programs in the national forests, the Forest Service often made allies out of the ranchers.

> A successful grazing program required forest officers to understand the local traditions of the people in their district. Officers tried to attend local livestock association meetings (to promote a right understanding of the purposes of forest reserves). These associations in turn sought official recognition from the Forest Service as described under Regulation No. 45 of the 1910 "Use Book" (grazing edition).[11]

CASE LAW DEVELOPS

When it came to legal ammunition to maintain control over grazing, the Forest Service relied on specific cases they felt backed their position: *United States* v. *Tygh Valley Co*, September 26, 1896; *Dastervignes* v. *United States*,

March 2, 1903; and *Dent v. United States*, March 26, 1904; each dealt with the issue of sheep grazing on forest reserves in defiance of forest reserve rules. The courts upheld forest reserve rules in all three cases on the basis that the closely herded sheep were damaging the reserves. However, as Judge Ethelbert Ward wrote in his essay, "The Legal Aspect of the Grazing Problem," these were essentially civil cases "by injunction" to prevent the damage to property, "and would apply as well to the individual as to the United States. They are founded upon the law of the land, and do not depend on rules and regulations."[12]

These three cases cited by the Forest Service to support their authority over livestock grazing on the forest reserves never raised the question of pre-existing grazing rights. Prior appropriation of water based on livestock grazing as proof of beneficial use was not an issue here. The legal basis of the preemptive private property rights argument was not decided here because the question was not asked.

Another grazing case which had been decided in favor of the Forest Service and against Montana was *Shannon v. United States*.[13] Judge William H. Hunt of the U.S. District Court of Montana ruled that the Forest Service did not have to fence the forest reserves against non-permitted livestock. The decision said the stockman must fence his livestock to prevent them from trespassing on government property. This ruling handed down March 18, 1907, preceded the U.S. Supreme Court ruling in *Kansas v. Colorado* by three months.[14]

Thomas Shannon, a Montana rancher, deliberately opened a fence to allow his cattle to stray on the Little Belt Mountain Forest Reserve. Shannon did not assert a prior-existing right to graze. He did not raise the issue of ownership of stock water based on the prior appropriation doctrine. Shannon challenged the right of the federal government to regulate his livestock grazing on the forest reserve lands in conflict with state grazing laws and the police power of Montana over its own territory. Judge Hunt ruled for the federal government.

In the essay cited above, Colorado Judge Ethelbert Ward commented on Judge Hunt's decision:

> Aside from exclusive state police regulations, there is another reason why the United States, as a landowner, cannot, by rules and regulations, nullify state laws. It is that to these lands the federal government is no more than an individual land holder.... Hence the United States have no better rights than the individual land holder.... This, I think, shows the error of Judge Hunt's decision nullifying Montana cattle laws.[15]

Ward went on to challenge Judge Hunt's assertion that the Montana Enabling Act on entering the Union, "forever disclaiming all right and title to the unappropriated public lands," was a relinquishment of state sovereignty over those lands.[16]

> The enabling act is not a relinquishment of state sovereignty, it is only a relinquishment of claim to the title of the public lands; and a pledge not to tax nor to interfere with its primary disposal. Had the enabling act in

specific terms relinquished state sovereignty and jurisdiction over the public lands, such clause would be absolutely void and unconstitutional. The Supreme Court of the United States has so declared.[17]

The court's recognition of states' rights, in particular the pre-existing rights of stockmen through the prior appropriation doctrine, had significant influence on Forest Service policy. The grazing permit system employed by the Forest Service began to emphasize the holders of pre-existing rights to the exclusion of those without pre-existing rights. The case of *United States* v. *Grimaud* reflected the emphasis being given to holders of pre-existing range rights.[18] In November, 1907, Pierre Grimaud, along with his associate J. P. Carajous, was indicted for pasturing sheep in the Sierra Forest Reserve of southern California without a permit. Grimaud's argument rested on the implied license to graze on the public lands as acknowledged in *Buford* v. *Houtz*.

Grimaud contended that his implied license granted under *Buford* v. *Houtz* negated a need for a permit from the Forest Service or the requirement to pay a grazing tax.

The U.S. Supreme Court ruled against Grimaud on May 1, 1911. The court ruled that the forest reserves had been withdrawn from the public domain so that the implied license in *Buford* v. *Houtz* did not strictly apply.

At issue was the fact that Grimaud did not have a permit. Grimaud represented the "tramp" sheepmen of the period who usually had no home base and moved bands of sheep wherever the grazing conditions were the most favorable. Grimaud made no valid assertion of prior rights in the Sierra Forest Reserve. The court went on to say that the implied license, "was curtailed and qualified by congress, to the extent that such privilege should not be exercised in contravention of the rules and regulations."

PRIOR RIGHTS RECOGNIZED BY PERMIT

What were those rules and regulations? Grazing permits were to be issued to:

A. Those who owned adjacent ranch property (established ranchers in the reserves).

B. Those who owned nonadjacent ranch property and traditionally used the public forest ranges (those with prior rights).

C. For transient herders who could make no claim to local property ownership.[19]

The class C permits came last in preference; A and B permits usually took all of the allotted ranges.*

The action by the Forest Service was a textbook application of the rules for granting grazing permits according to the Forest Service "Use Book." It was a classic case of denying grazing privileges on the basis of the stockman having no pre-existing rights in the Sierra Forest Reserve. At the time of the Grimaud case there were extensive valid permits to graze in the Sierra Forest Reserve with which Grimaud's grazing activities would have conflicted.

The essence of the ruling in *United States* v. *Grimaud* was that the implied

Cf. supra, pp. 155-56.

license mentioned in *Buford* v. *Houtz* had been "curtailed and qualified" in favor of those stockmen who could prove they had pre-existing grazing rights on the Sierra Forest Reserve.

A companion case, *Light* v. *United States*, was ruled on at the same time by the U.S. Supreme Court on May 1, 1911.[20]

Fred Light was a rancher grazing cattle in the Holy Cross Forest Reserve in Colorado. Light and his neighboring ranchers were concerned that the Forest Service tax on grazing could become prohibitive. They felt if the tax were allowed to stand the Forest Service could eventually raise the tax until it priced the stockmen off their range. The Light case was a test of the Forest Service grazing fee. It was conducted in an almost friendly manner. Light agreed to put his cattle on the reserve without paying a grazing fee or applying for a permit. He did not demand the recognition of his prior right to graze through the permit system, but deliberately chose to assert his rights through the implied license clause of *Buford* v. *Houtz*. The court ruled that the "implied license" had been "curtailed and qualified" by Congress, to the extent that such privilege should not be exercised in contravention of the rules and regulations. *The rules and regulations referred to here were based on giving preference to those who had their prior rights formally recognized through the permit system.*

On May 24, 1897 the U.S. Supreme Court had rendered a decision in a case where a stockman by the name of Daniel A. Camfield had fenced a large portion of federal land by taking advantage of the even-odd ownership pattern of land adjacent to a railroad in Colorado. Camfield had purchased the odd numbered sections and fenced the entire township including the even numbered sections owned by the federal government through a careful survey which allowed him to keep his fences entirely on his private land yet prevent public access to the federal land.[21]

The federal government successfully argued that the Act of February 25, 1885, "An Act to Prevent Unlawful Occupation of the Public Lands" allowed the government to force the removal of Camfield's fences. The government argued that Camfield's fences reduced the sale value of the public lands because the public was denied access.

In the Camfield case, as in *Tygh Valley*, *Dastervignes*, and *Dent*, the issue of stock water on the federal lands based on livestock grazing as proof of beneficial use, which had been established prior to March 3, 1891, was not raised.[22] The preemptive private rights argument continued to lie dormant as far as a legal challenge was concerned. The decisions in the cases of Grimaud, Light, and Shannon all indirectly supported the preemptive rights argument. Each case supported the permit holder over the non-permit holder. The permit holder was the one who could prove a prior right to graze in the forests.

ECONOMIC ATTACK ON PRIOR RIGHTS

While some western ranchers were benefiting by their preference and privileged position on the national forests, there was an underlying fear in the minds of many ranchers that failure to recognize definitive private property rights to the forest ranges would eventually lead to instability.

In 1905, stockmen suspected that the government had only grudgingly admitted the importance of grazing. Many believed their privileges would be short-lived, fearing that the managers of the forests would become more oriented toward the single purpose of timber management, especially when the professional foresters from the eastern schools obtained secure control over the forest reserves.[23]

This was the concern expressed by Fred Light when he agreed to test the Forest Service grazing fee concept. The fee was continually referred to as a tax by stockmen. *Ranch and Range*, a stockmen's publication, referred to the fees as "taxation without representation."[24]

By 1916 the worst fears of the stockmen appeared to be materializing. The western stockmen who had originally been recruited by the forest reserve proponents to administer the grazing program were being replaced by non-stockmen, foresters who were more in tune with the original objectives of northern core interests.

The Forest Service was, in turn, alarmed over widespread recognition of the rancher's grazing rights on the forests as real property. Bankers loaned money on the collateral value of the entire ranch including range rights on the forest. When ranches were sold the range and water rights were a part of the ranch sale. Children inherited the range rights and water rights from their parents. Courts continued to recognize the real property value of range and water rights. Forest administrators responded with a policy of discouraging private investment in range improvements. This policy was viewed as preventing the strengthening of private preemptive rights claims. Ranchers were encouraged to ask the Forest Service to build improvements or at least have the Forest Service supply the material. The Forest Service viewed the latter approach as increasing the "public's" share in the split estate.

If the preemptive rights claim could not be erased or contained by current methods some felt it was time to divest the rancher of his preemptive rights by pricing him off the range. Fee increases occurred from 1909-1912 and again in the 1915-16 period. A campaign to raise the grazing fee to a "commercial basis" was begun in 1917. After a storm of protest by the stockmen grazing fees were increased in 1919. Increased stocking of the ranges was allowed to help quiet the protests over the fee increase.

The year 1919 brought out many of the arguments for fee increases which the livestock industry and the national forest proponents would be arguing about seventy years later.

"Low fees cause overgrazing."

"Fees should be raised to market value for the privilege."

"An immediate 300 percent increase should be imposed."

All these arguments to raise the grazing fees to prohibitive levels came at a time when the Forest Service was fighting a losing battle to keep permits from assuming value in the commercial grazing community. The Forest Service did not wish to officially recognize their values because the holder might then claim a permanent possessory right over the use of the forest range.[25]

It was during this period, 1919-20, that the grazing fee formula was first proposed. Albert F. Potter retired from the Forest Service in 1920. This signaled an end to the power of the grazing interests in the Forest Service. The grazing fee formula proposed at this time reflected the conflict between those who wanted to price the stockmen off the range and those like Potter who understood the rancher's private investment in the range. The grazing fee formula was an attempt to tie the grazing fee to a recognition of the rancher's cost on the forest range.

Efforts were underway to raise the grazing fee to equal the lease rate on private pasture. Men like Potter understood that a private pasture lease was worth much more to the rancher because the private landlord provided virtually all of the necessary production inputs. The private landlord lessor provided almost all of the cost of use. A private lease was worth more because the lesee stockman was receiving more. On the forest range, by comparison, the rancher himself was providing most of the production inputs plus all of the cost of use. To compare the forest range with private pasture for lease rate purposes was like comparing apples to oranges. Potter understood this. The anti-livestock movement supporting the Forest Service also understood this. They understood that if they could sell this "fair market value" argument to the public, Congress might pass supportive legislation. In that case the stockman, with his claim to preemptive property rights in the forest, would be disenfranchised economically.

The controversy over preemptive rights and grazing fees developed into a showdown in Congress over some pro-stockmen grazing bills introduced in 1925-26 and corresponding committee hearings throughout the West in the spring and fall of 1925. The committee had generated considerable publicity about the bills that tended to be generally negative to the Forest Service. The Forest Service struck back through their friends in the eastern conservationist press. The American Forestry Society's periodical—then called *American Forests and Forestry*—continued to publish conservationist views that were widely disseminated by the eastern news media. The *New York Times* and *Saturday Evening Post* took the stockmen to task.

Chief Forester Greeley made a statement in a *Saturday Evening Post* article that illustrated the Forest Service's contradictory position of denying the existence of the stockmen's property rights in the range while admitting those rights had been recognized.

Greeley "referred to the handful of stockmen and their leaders who were attacking the basic policy that governed the forest ranges. He condemned the demand by stockmen that grazing rights be confirmed by laws as an attempt to secure special privilege for a few users who would then have the right to sell these privileges at high prices. Greeley pointed out that the Department of Agriculture had always recognized the rights of prior users of grazing lands. 'It has never, however,' he declared, 'been able to sanction the claim to a vested right of property interest in the forage. On the contrary, it has regarded that conception as absolutely inimical to the principles of conservation.'"[26]

Even though the Forest Service had been forced, early on, to "recognize the rights of prior users of grazing lands," it had never abandoned the objectives of being able to grant permits at will to all applicant graziers who would owe their privilege strictly to a "gift" or "grant" of the federal government. These types of permits to graziers with no claim to prior rights would be revokable at will by the Forest Service. These types of permits would not raise the prior rights argument.

The prior rights holder was continually defending his preference from the demands of other stockmen attempting to gain an interest in the forests. This ongoing conflict between those who had pre-existing rights in the forests and those who were seeking to establish new claims would result in some very express language protecting prior rights in the next major piece of land use legislation involving the federal lands. Stockmen in the early 1930s would demand that the Taylor Grazing Act fully and expressly protect all prior rights in the land and expressly prohibit the acquiring of rights in the land solely by the creation of a grazing district or the granting of a grazing permit by the federal government.

"SCIENTIFIC" RANGE MANAGEMENT

By 1910 the Forest Service had begun a program of range "data" collection that would be used in the future to justify a reduction of livestock on the ranges as well as fee increases. Scientific range management was emerging as a tool which would be used successfully over the years to increase Forest Service budget demands and to combat the stockman and his preemptive property rights claims.

The Organic Act of June 4, 1897 gave the Forest Service the authority to "improve and protect the forests." From this time on, scientific range management would develop as the tool by which political objectives would be reached. All moves against the stockmen would be couched in terms of resource damage. As the Forest Service budget grew, the amount of public money available to expand the range data bank would also grow. This range data was developed with the expenditure of public money to use against the stockman who did not have public money to spend on range data. It would be used to increasingly intimidate the ranchers and quiet their claims to property rights.

The resource damage argument would accomplish Forest Service objectives without raising the dangerous issues of economic damage and the taking of private property without compensation. The Forest Service would be able to economically damage and take the property of uncooperative stockmen without raising the Fifth Amendment of the U. S. Constitution challenge with its guarantees of due process of law and just compensation.

MANY CHANGES

By the 1930s competition for control of the remaining federal rangeland was well underway. The Forest Service had been vying for control since 1905. Several bills to transfer all of the West's grazing land to the Forest Service had failed.

The Hoover administration brought with it a program to privatize federal range lands. The Forest Service's northern core political supporters panicked.

The Forest Service was, by 1929, a rapidly growing federal agency whose annual appropriations had grown from less than a million dollars in 1905 to fifteen million dollars by 1929. Hoover's attempt to recognize private rights in the forests was a threat to the salaries, budgets, careers, and power of the Forest Service.

The proposed privatization of western rangelands under President Hoover touched off zealous efforts on the part of the Forest Service and the General Land Office of the Interior Department to protect their own proprietary interests as administering agencies.

The most vigorous efforts came with the election in 1932 of Franklin D. Roosevelt and his appointment of Harold L. Ickes as Interior Secretary. Ickes embarked upon a campaign to have the national forest administration moved out of Agriculture and back under control of the Interior Department. The Forest Service responded with harsh criticism of the land management policies of the General Land Office. Forest Service criticism emerged formally in a report to Congress in 1936 titled, "A Report on the Western Range: A Great but Neglected Natural Resource."

The report emerged after jurisdiction of the remaining federal rangelands had been consolidated in Interior under the Taylor Grazing Act of 1934.

The Western Range or *Green Book*, as it was often called, was the product of Associate Forester Earle Clapp. The tone of the report is illustrated by the following quote: "There is perhaps no darker chapter nor greater tragedy in the history of land occupancy and use in the United States than the story of the Western Range."[27]

Critics labelled *The Western Range* a propaganda document: it had only praise for Forest Service range administration and sounded like "sour grapes." This latter referred to the Forest Service's failure to gain control of rangelands now managed by the Interior Department under the Taylor Grazing Act.[28]

Others noted the Forest Service's "self righteous" criticism of other land agencies and overrated analysis of their own accomplishments. The Forest Service critique was attacked by F. C. Mollin, secretary of the American National Livestock Association as a "self-serving compilation of falsehoods."

The Western Range further proposed legislation regarding the federal rangelands.

1. Equitable distribution of grazing privileges.

2. Reduction to bring down numbers of stock on the range.

3. Consolidations of range administration under one federal department.

4. Avoidance of any move by users to establish preemptive or property rights to the range.

5. Revision of the grazing act "in the public interest" to prevent a conflict between federal and state authority.[29]

Point one proposed the long held objective to cloud or remove the rights of the original forest permittees whose permits had been granted on the basis of prior rights, in favor of those "new" permittees who would hold their permits solely as a grant from the federal government.

Point two urged a reduction in the stockman's livestock. As the livestock units were reduced the cost per unit increased (economic law of diminishing returns). At some point in time it becomes economically unfeasible to run livestock on the range. This approach accomplished the infringement or loss of the stockmen's property values in the range. The Fifth Amendment issue was not raised as long as the reason for the reduction could be "to avoid range damage." Range damage could be invoked at will by the Forest Service which exclusively and with public monies conducted range studies, analyzed those studies, and interpreted those studies to accomplish their own objectives.

Point three, urging consolidation of range administration under one federal department, was clearly an effort to expand the Forest Service. The Forest Service was on record since 1905 as trying to accomplish that objective.

Point four focused on the ever-lingering possibility that the prior rights claims of the stockmen might be raised in such a way that the courts and Congress would have to recognize them formally. Formal recognition of the preemptive rights of stockmen was considered mutually exclusive of the survival of the land management agencies.

Point five targeted "the grazing act." The law referred to was the Taylor Grazing Act of 1934. Forty years of experience by stockmen dealing with the 1891 Forest Reserve Act and its attempts to deny property rights in the federal lands acquired under state law caused stockmen and western state interests to demand express protection of those rights in legislation relating to non-national forest grazing rights.

The act was specific in recognizing water rights on federal lands. It specifically recognized private investment in range improvements as well as rights of ways and all other prior existing rights.

The act was also specific in barring federal agents from creating their own clientele among graziers in the grazing districts by granting permits or including within a grazing district those stockmen without prior rights. The last line of Section 3 states, "The granting of a permit or the creation of a grazing district pursuant to this act creates no right, title or interest in the land."[30]

Chapter Nine Footnotes

1. *United States Constitution,* Article I, Sec. 8, Clause 17.
2. *Pueblo Chieftain,* May 15, 1907. Forest Service Law Office Correspondence, Record Group 49, Drawer 16, National Archives.
3. Ibid.
4. W. J. Palmer to Pinchot, May 23, 1907, Forest Service Law Office Correspondence, Record Group 49, Drawer 16, National Archives.
5. W. J. Palmer to P. P. Wells, (No date), Forest Service Law Office Correspondence, Record Group 49, Drawer 16, National Archives.
6. G. S. Arnold to T. E. Will, June 24, 1907, Forest Service Law Office Correspondence, Record Group 49, Drawer 16, National Archives.
7. P. P. Wells to T. E. Will, June 29, 1907, Forest Service Law Office Correspondence, Record Group 49, Drawer 16, National Archives.
8. Ibid.
9. William B. Greeley, "The Stockmen and the National Forests," *Saturday Evening Post,* November 14, 1925, pp. 10-11, 80, 82, 84.
10. Forest Service. U. S. Department of Agriculture, *The Use of the National Forest Reserves,* Washington, D.C., July, 1905, edition 1910. Known as the "Use Book."
11. Rowley, *U.S. Forest Service Grazing and Rangelands: A History,* p. 80.
12. Forest Service Law Office Correspondence RG 49, Drawer 16, National Archives.
13. *Shannon* v. *United States,* 160 F. 870 (9th Cir. 1908).
14. *Memorandum Opinion,* No. 725, Circuit Court of the United States, Ninth Circuit, District of Montana, in Forest Service Law Office Correspondence RG 49, Drawer 16, National Archives.
15. "The Legal Aspects of the Grazing Problem, by Ethelbert Ward, Denver, Colorado," Forest Service Law Office Correspondence RG 49, Drawer 16, National Archives.
16. Ibid.
17. Ibid, citing *Pollard* v. *Hagen,* Ala., 3 HOW. 223, 11 L Ed. 565.
18. *United States* v. *Grimaud,* 220 U.S. 506, 31 S.Ct. 480, 55 L.Ed. 563 (1911).
19. Rowley, *U.S. Forest Service Grazing and Rangelands: A History,* pp. 58-59.
20. *Light* v. *United States,* 220 U.S. 523, 55 L.Ed. 570, 31 Sup. Ct. Rep. 485. (1911).
21. *Camfield* v. *United States,* 167 U. S. 518. (1896).
22. *United States* v. *Tygh Valley Co.,* 76 F. 693 (1896); *Dastervignes* v. *United States,* 122 F. 30 (1903); *Dent* v. *United States,* 76 P. 455, reversing 71 P. 920 (1902).
23. Rowley, *U.S. Forest Service Grazing and Rangelands: A History,* p. 57.
24. *Ranch and Range,* March 1907, as quoted in Jenks Cameron, *The Development of Government Forest Control in the United States,* Johns Hopkins University Press, Baltimore, 1928, pp. 242-43.
25. Cameron, *The Development of Government Forest Control in the United States,* pp. 333-34.
26. William B. Greeley, "The Stockmen and the National Forests," *Saturday Evening Post,* November 14, 1925, pp. 10-11, 80, 82, 84. George F. Authier, "Both Sides of the Range Controversy," *American Forests and Forest Life* Vol. 31, December, 1925, pp. 715-170.
27. *U.S. Congress,* Senate, "A Report on the Western Range: A Great But Neglected Natural Resource," 74th Congress 2d Sess. 1936, *S. Doc.* 199, p.3.
28. Rowley, *U.S. Forest Service Grazing and Rangelands: A History,* p. 156-7.

10

PRIOR RIGHTS AND THE TAYLOR GRAZING ACT

The Taylor Grazing Act of June 28, 1934, applied to most federal grazing lands not then covered by the Forest Reserve Act. For the most part they were more arid lands and lands of lower elevation administered by the Department of the Interior. The Forest Service and its northern core proponents had worked diligently since 1905 to wrest control of these lands from the rival agency. The Interior Department strongly resisted the Forest Service takeover attempt—particularly after the appointment of Interior Secretary Harold L. Ickes in 1933. Secretary Ickes wanted growth in the Interior Department comparable to the Forest Service. Obviously, this could not be accomplished by giving away administration of the land to another agency. Neither could Interior Department growth be accomplished by recognizing private title to the land.

The ranching community had learned from the experience with the forest reserves. The states had also learned. The western states had developed extensive statutory water law since the 1890s to strengthen control of water on the federal lands within their borders. No longer were claimants dependent on living witnesses alone to consolidate their claims to vested rights. State statutes now offered a procedural approach for confirming prior appropriation. Stockmen had availed themselves of the procedures set forth in state statutes and virtually no water on the federal lands was unappropriated by 1934.

Stockmen in the early 1930s felt they had an ownership interest in the range. The Hoover administration had made an effort in 1929 to recognize the stockmen's rights and privatize the western rangelands. Both the Interior Department and the Forest Service opposed privatization for their own proprietary reasons. Privatization would eliminate much of the reason for the agencies' existence. The northern core was particularly opposed to privatization, if for no other reason than recognition of the stockmen's preemptive rights on the non-forest reserve federal lands automatically breathed life into the preemptive right claims on the forest reserves. To recognize those claims in one land category meant the recognition of those claims in the other.

By this time—1934—the forest reserves had been in place for as much as forty-three years. Vast resource wealth had been stripped from the hands of the western settlers and the western states—those who believed they had a legal claim to it through the prior appropriation doctrine. The prior appropriation doctrine had been upheld as had the states' sovereignty over the forest reserves in *Kansas* v. *Colorado*. But the courts had also upheld the right of the Forest Service to regulate in defiance of state law in *Shannon* v. *United States* and *Light* v. *United States*, resulting in inadvertently formalizing a split estate ownership of the land and water between private parties and the federal government. The implications of this confused state of affairs were grave. Any move at this late date that could force recognition of the original preemptive right claims would automatically result in huge damage claims against the federal government similar to the multimillion dollar claims successfully made by various Indian tribes in later years.

To avoid opening the Pandora's box of prior rights with its potential for unparalleled damage claims, Congress passed a law recognizing and protecting the prior right claims of stockmen short of fee title. It created grazing districts and granted permits in the non-forest federal rangelands.

Congressman Edward T. Taylor of Colorado had a background close to the livestock industry. He had been a major opponent of the forest reserves and a strong proponent of privatization. Faced with the reality of northern core resistance to any action that would expose the questionable legality of the forest reserves, Taylor opted for what he thought would be a bill adequately protecting the rancher's property rights. The bill contained a clause that provided for the final disposition of these lands. Many stockmen viewed the term "final disposition" as being synonymous with eventual recognition of preemptive rights.

Taylor and the livestock community emphasized language protecting the stockmen's preemptive rights pending final disposition. In Section I we read:

> Nothing in this sub-chapter shall be construed in any way to diminish, restrict, or impair any right which has been heretofore or may be hereafter initiated under existing law validly affecting the public lands, and which is maintained pursuant to such law except as otherwise expressly provided in this sub-chapter.[1]

In Section III the act says:

> That nothing in this sub-chapter shall be construed or administered in any way to diminish or impair any right to the possession and use of water for mining, agriculture, manufacturing, or other purposes which has heretofore vested or accrued under existing law validly affecting the public lands or which may be hereafter initiated or acquired and maintained in accordance with such law.[2]

In Section IV we find a statement that the stockmen demanded in order to protect their private capital investment in range improvements. By 1934 the rangelands of the West were dotted with wells, livestock handling facilities, living quarters, roads, fences, crossfences, pipelines, reservoirs, etc., representing millions of dollars of private capital investment. These and future private investments needed specific protection as private property.

> No permit shall be issued which shall entitle the permittee to the use of such improvements constructed and owned by a prior occupant until the applicant has paid to such prior occupant the reasonable value of such improvements.[3]

State governments were anxious to avoid the attempts to deny the sovereign rights of the states as occurred with the first forest reserve withdrawals until the

U. S. Supreme Court in the case of *Kansas* v. *Colorado* specifically upheld states' rights. Section I addresses this concern with the appropriately-placed phrase, ". . . nor as limiting or restricting the power or authority of any State as to matters within its jurisdiction."[4]

Along with these statements protecting the prior rights of individuals and the states the last line of Section III says: "but the creation of a grazing district or the issuance of a permit pursuant to the provisions of this sub-chapter shall not create any right, title, interest, or estate in or to the lands."[5]

THE REAL MEANING OF SECTION III

This last line has been interpreted to mean that the stockmen had no right, title, interest, or estate in or to the land: but that is not what it says. It plainly says that the *creation of a grazing district or the issuance of a permit shall not create. . . .* If, as some would argue, this phrase negated all prior rights, then the specific language that precedes and follows this phrase would be superfluous. Clearly the bill protects *all prior rights.*

And what were those rights? They were rights acquired under the prior appropriation doctrine: water and the use of the land the water served. What was the legal authority for those rights? Local law and custom and decisions of the court. This authority was clearly stated in the Act of July 26, 1866. As the U.S. Supreme Court affirmed in *Broder* v. *Water Company*, this Act constituted "a voluntary recognition of a pre-existing right of possession."[6]

The prior rights protected by the Taylor Grazing Act were water rights and associated property acquired under the prior appropriation doctrine. These were livestock waters and their associated facilities. The proof of beneficial use, which local law and custom and court decisions recognized, was livestock grazing. The stockmen at the time of the passage of the Taylor Grazing Act believed they already had a certain right, title, interest, or estate in the land. Until those claims were clarified and until final disposition of the land, they demanded and received statutory protection for those rights.

The Senate hearings on the Taylor Grazing Act addressed this very topic. Interior Secretary Ickes recognized these claims in soliciting the stockmen's support for the Taylor Grazing Act that would put the non-forest grazing lands under Ickes' jurisdiction. In exchange for the rancher's political backing Ickes promised to formalize the existing range rights in the assignment of permits. In the hearings Ickes said, "We have no intention to . . . drive stockmen off their ranges or deprive them of *rights to which they are entitled either under State laws or by customary usage.*"[7] [My italics.]

The language in the last line of Section III was demanded by the stockmen. Experience with the Forest Service was fresh in their minds. The Forest Service, from the beginning of their grazing program, had granted some grazing permits to parties with no prior rights. These permittees often competed to the detriment of the permittees with prior rights or preference. They had sometimes attempted to establish rights by use on the water and range of the preference permittee. Expensive and bitter litigation had often been the result. To protect themselves from any attempts by the administrators of the Taylor Grazing Act to build a grazing clientele who owed their grazing privileges strictly to the government, the stockmen supported the language: "the creation of a grazing district or the issuance of a permit pursuant to the provisions of this sub-chapter shall not create any right, title, interest, or estate in or to the lands."[8]

In 1934 the overlapping claims of multiple stockmen to the same range and water needed clarification. An anti-fencing law from 1885, which had been enforced to varying degrees since the first term of Grover Cleveland, had compelled many stockmen to run livestock in common. Not only did this contribute to overuse of some range, it also prevented the clear definition of property rights. It was more important in the minds of Interior and Forest Service officials to prevent definition of property rights than to protect the range resource. Thus began the long period of range abuse that characterized that period of federal range control. Common-use conflicts have been worked out between the ranchers in subsequent years. Range-line agreements creating individual fenced range allotments are now predominant throughout the West. These individual allotments clearly define property rights.

In 1934 there was much competition between stockmen who ran livestock in common on the same range. Until those conflicts were settled, stockmen wanted assurance that permit applicants would not gain rights or lose rights as a result of the time of application, favoritism from the agency, or other extenuating circumstances. The no "right, title, interest, or estate in or to the land," language was supported by stockmen to protect themselves from built-in conflicts resulting from common range.

The Interior Department supported the same language in the Act because it wanted to discourage any move to strengthen existing private rights in the split estate.

LOCATION OF STOCKWATER USE

Some have tried to explain valid pre-existing stock water rights on the federal lands without recognizing a valid pre-existing right to graze. Their argument usually centers around the most common type of grazing permit; the permit which is commensurable to private lands.

The act states in Section III:

> Preference shall be given in the issuance of grazing permits to those within or near a district who are landowners engaged in the livestock business, bona-fide occupants or settlers, or owners of water or water rights, as may be necessary to permit the proper use of the lands, water or water rights owned, occupied or leased by them.[9]

Those "who are landowners" are those with commensurable private property.

Those "who own water or water rights" are just that—those who own water or water rights.

To support the "no pre-existing grazing right argument," some say that the water rights are only valid as stock water rights when they flow on to the commensurable private lands. Therefore livestock grazing—which is proof of beneficial use—occurs on private, not federal land. The weakness in this argument is obvious. Many valid stock waters do not flow to private land. Many valid stock waters are springs, seeps, or wells where use of the water by livestock is localized around the water, often many miles removed from the nearest private lands.

Land based permits were established on the premise that a certain amount of private land was associated with ranch operations. It was to this piece of private commensurable land that the grazing permit was attached for the purpose of legal reference. The range, which was acknowledged by the permit, would in some cases extend as far as one hundred miles from the commensurable private property. Within that distance would be many private waters and livestock using those waters. Livestock grazing, as proof of beneficial use, was miles removed from the land base to which the permit was attached.

Even stronger evidence that pre-existing grazing rights are the prior rights being protected by the Act, is the second major type of permit issued pursuant to the Taylor Grazing Act. It is called a water base permit. Section III not only referred to "landowners" but to the "owners of Water or Water rights." These were stockmen, usually sheepmen, who owned no base private land. They moved with their herds or flocks from water to water as available forage allowed, living out of camps. Many desert livestock operations were water based. The grazing permit referenced individual waters or water rights owned by the stockmen that arise on the federal lands as a legal reference for the permit. These water base permits also clearly identify an area in a specific radius of a given water as the grazing area associated with the permit. Neither the water nor the grazing area is on private ground. Usually these water based grazing rights are miles removed from the nearest private land. The right to these waters was established under the prior appropriation doctrine during the early settlement of the West. Most of the priorities are no later than the 1880s. The proof of beneficial use acknowledged by local law and custom and court decisions was livestock grazing. All of this occurred before the General Land Law Revision Act of 1891 with its forest reserve clause and preemption repeal clause. This was established before the Taylor Grazing Act. Clearly the water base permit poses an unsolvable enigma to those who would argue that the Taylor Grazing Act does not protect a pre-existing grazing right. That is the reason water base permits have been the target of agency attack more than land base permits.

CASE LAW AND THE TAYLOR GRAZING ACT
The discussions of pre-existing grazing rights cannot be complete without a review of the major case law involving the Taylor Grazing Act.

Proponents of the "no pre-existing grazing right" argument cite case law showing a grazing permit as a mere privilege granted by the government that can be revoked or withdrawn without payment or compensation.

They cite holdings from the following cases, condensed here in key summaries by the editors of the court report documents:

Mollohan v. *Gray*—Annual grazing permits authorizing owners to use portions of federal range pursuant to the Taylor Grazing Act conferred upon recipients a mere privilege to graze livestock, and this privilege could be withdrawn by the United States without payment of compensation.[10]

Acton v. *United States*—Grazing permits issued pursuant to the Taylor Grazing Act created no interest or estate in public lands but only a privilege which might be withdrawn and no property rights would accrue to the licensee upon revocation which would be compensable in condemnation.[11]

United States v. *Cox*—Permits issued to ranchers granting exclusive or preferential rights to graze stipulated number of cattle on public domain although valuable to ranchers, were not interests protected by the Constitution against taking by the Government, which granted them with understanding that they could be withdrawn at any time without payment of compensation.[12]

Oman v. *United States*—Grazing permits are only a privilege withdrawable at any time for any use by the sovereign without payment of compensation.[13]

Bowman v. *Udall*—Statutory grazing permits and leases confer a mere privilege to graze livestock which can be withdrawn by the United States without payment or compensation.[14]

Placer County Water Agency v. Jonas—holder of permit . . . in which holder agreed that rights he would receive would give way to any use or need of government, did not have compensable interests in such property.[15]

Fauske v. *Dean*—Permits . . . are mere licenses which are revokable at will without legal right to compel compensation.[16]

United States v. *Fuller*—The provisions of the Taylor Grazing Act . . . make clear the congressional intent that no compensable property be created in the permit lands themselves as a result of the issuance of the permit.[17]

These court decisions are often presented as proof that no pre-existing grazing rights were protected by the Taylor Grazing Act. But each of these cases plainly discuss grazing *permits* as revokable and non-compensable. They just as plainly do not mention water rights or in any way reference the proof of beneficial use for those waters. They do not discuss the prior appropriation doctrine or any rights that have "vested or accrued under existing law validly affecting the public lands."

Under the Taylor Grazing Act, as well as under national forest regulation, permits were granted to those who could prove they had valid prior rights. To obtain a permit under the Taylor Grazing Act a stockman had to prove he had priority year use of the range in question. He had to prove he had commensurable private property or sufficient water base. His ownership of water or water rights was his strongest claim to a range area. To obtain a permit the stockman had to prove that he owned prior-existing rights. These were rights created under the prior appropriation doctrine of the states. These rights were not created by the federal government and were clearly not revokable by the federal government. Instead, the federal law clearly protects those private rights in the federal lands.

Case law exists on this topic also. In *Faulkner* v. *Watt*, the court held that the purpose of the Taylor Grazing Act was to stabilize the livestock industry and protect the rights of sheep and cattle growers from interference.[18]

The court held similarly in *Barton* v. *United States* that the Taylor Grazing Act was designed to stabilize, preserve and protect the use of public lands for livestock grazing purposes including the extent to which the lands may be grazed.[19]

In *Chournos* v. *United States*, the Tenth Circuit Court of Appeals held that the purpose of the Taylor Grazing Act, in authorizing issuance of grazing permits to full carrying capacity of public lands to owners of base lands classified as dependent by use on such public lands, is to stabilize the livestock industry and to permit use of public range according to the needs and qualifications of livestock operators with base holdings.[20]

The United States Court of Appeals for the District of Columbia in the case of *Red Canyon Sheep Co.* v. *Ickes* held that the purpose of the Taylor Grazing Act was to provide for the most beneficial use possible of public range in the interest of grazers and the public at large, to define grazing rights and to protect those rights by regulation against interference.[21]

In these latter cases the court discusses grazing rights, base lands, and rights in general. When referring to grazing permits in these cases, the reference is to the rights that the stockmen owned upon which the permit was based.

THE RIGHT VERSUS THE PERMIT

Under the Taylor Grazing Act, as under the forest regulations, the permit acknowledged the pre-existing right and set a value on it measured in animal unit months (AUM's) for the purpose of assessing a fee. The granting of a permit did not create any rights. It acknowledged pre-existing rights and set their value. The permit was created by the federal government and was revokable. The underlying property rights upon which the permit was based were not created by the federal government and were not revokable. Such is the dilemma surrounding the split estate rangelands of the western United States. Containment of the underlying property rights issue has primarily been accomplished by federal land management agencies by settling threatening disputes before they reach the federal courts.

FEDERAL LANDS AND WELFARE

The Taylor Grazing Act developed in the midst of the Great Depression. It was passed at a time when the United States was turning to the left politically.

The federal government under the administration of Franklin D. Roosevelt involved itself in the private and business affairs of Americans in an unprecedented manner. Bureaucracy and bureaucrats, once the target of disdain and ridicule, were gaining acceptance. Federal involvement in income redistribution schemes proliferated. The Forest Service expanded its influence and budget by becoming a joint manager of the Civilian Conservation Corps (CCC) program with the Army in 1933. Harold Ickes, Secretary of Interior, elicited stockmen's support for the Taylor Grazing Act with promises of federal expenditures on range improvements if they supported his version of the legislation.[22] Ickes was appointed head of the Public Works Administration (PWA) in 1933. A new dimension in federal land management was born in the Great Depression. The federal bureaucracy became a major source of employment. The bureaucracy geared up to absorb the unemployed and the unemployable. Record unemployment rates (11 million in 1934) triggered more intensive management practices on forestlands and lands under jurisdiction of the Taylor Grazing Act. The agencies and the number of their employees ballooned under this expanded mandate.

The massive growth and proliferation of federal agencies during the New Deal of the 1930s was arrested by World War II. But as the war approached its conclusion, political consideration of post war unemployment emerged. The Truman administration, considering the political implications of high unemployment from the return of servicemen and the turndown in the defense industries, groped for a solution.

To resume the direct welfare programs of the New Deal would be a Democratic admission that the New Deal had not solved the Depression. Certainly this could only help the Republicans in the coming congressional elections. A solution was found in the Employment Act of 1946.[23]

The Employment Act of 1946 opened the door for the expansion of existing government agencies and the creation of new federal agencies. The declaration that opens the act said: "The Congress declares that it is the continuing policy and responsibility of the Federal Government to use all practicable means consistent with its needs and obligations and other essential considerations of national policy, with the assistance and cooperation of industry, agriculture, labor, and State and local governments, to coordinate and utilize all its plans, functions, and resources for the purpose of creating and maintaining, in a manner calculated to foster and promote free competitive enterprise and the general welfare, conditions under which there will be afforded useful employment opportunities, including self-employment, for those able, willing, and seeking to work, and to promote maximum employment, production, and purchasing power."[24]

The key to understanding this long-winded declaration is the final phrase, "to promote maximum employment, production, and purchasing power." No longer would participants in government work programs be considered welfare recipients. Those absorbed into federal agencies under the Employment Act of 1946 would be given government service ratings. Agency hiring of the unemployed after World War II gave preference to ex-servicemen. The Forest

Service expanded in accord with the Employment Act of 1946. In addition, in 1946 the old General Land Office was combined with the Taylor Grazing Office to form a new and larger agency called the Bureau of Land Management (BLM).

The *make-work* philosophy of the New Deal was now formally ensconced in government service. Most of the ex-servicemen absorbed by the federal agencies in this post World War period had developed their practical economic understanding under the welfare philosophies of the New Deal and the *cost plus* government spending policies necessitated by war conditions. A philosophy based in the belief that the United States Treasury was an inexhaustible source of taxpayer's money led to the profligate spending practices of the agencies. These spending practices were punctuated with ongoing demands for budget increases. As long as the New Deal philosophy significantly influenced Congress, Congress was disposed to pass legislation creating programs that, though non-cost effective in themselves, were justified on the grounds that the programs would create "meaningful employment."

The Multiple-Use, Sustained-Yield Act of 1960 was promoted on the basis that federal land management should not be single-use oriented.[25] The *single use* orientation of the BLM toward grazing as directed by the Taylor Grazing Act was a prime target. The Multiple-Use, Sustained-Yield Act expanded agency concerns to specific areas such as watershed protection and wildlife habitat management. Multiple use had always been a fact on the western rangelands. Good range management was good watershed management. Good range management was good wildlife habitat management. Recreation, hunting, and fishing were a protected fact of life on the federal lands.

The Multiple-Use, Sustained-Yield Act helped satisfy two major goals of federal land management in the West: (1) it increased the budgets, salaries, and careers of the affected agencies and (2) it formalized the claims of other multiple use interests. These claims could now be used to assault the stockmen's hold on his rangelands, specifically his water rights acquired under the prior appropriation laws of the states. The *concerns* and *needs* of the newly created client groups would be used to create resource argument attacks against the stockmen on the federal lands of the West. Watershed protection arguments would be used to justify livestock reductions. Wildlife habitat protection would be used to justify livestock reductions. Recreationist needs would be used to justify livestock reductions.

Use of the federal lands for political and public welfare purposes was not new.* At the close of the Revolutionary War, Congress made land on the federal domain available to veterans who would go West and settle. A similar action was taken after the War of 1812, the war with Mexico in 1845, and the Civil War of the 1860s. Placing veterans on the nation's frontier not only defused troublesome unemployment problems in the East, but created a vanguard of quasi-military settlers on the nation's frontier.

Congressional action was taken to encourage homesteading in the West after the Spanish-American War of 1896, but by this time virtually all the water on the federal lands had been claimed. Forest reserves were covering some of

*Supra, p. 32-34.

the most desirable agricultural acreages. Claims to range rights, mining claims, right of ways, easements, and a host of other lawful property interest on the federal lands severely limited the amount of land which could be effectively homesteaded for farming.

Efforts to encourage homesteading by veterans after World War I was a dismal failure despite amendments to the Homestead Act such as the Stock Raising Homestead Act. Two major factors were in play here. By 1918 it was very apparent that the federal lands of the West had been covered with so many claims that little, if any, federal land was available without some cloud on the title. The federal lands were no longer public lands, owned solely by the federal government, but split estate lands where most, if not all, of the utility values were controlled by private parties. The second reason homesteading failed to appeal to World War I veterans was the dramatic change from a rural to an industrial society that was underway in the nation. A well-paying job in America's urban industrial centers presented a more enticing future than the hardscrabble life of a pioneer in the arid West.

When the approaching end of World War II raised the specter of politically unacceptable unemployment rates, the federal government once more looked to the federal lands for a solution. The federal lands played a role, but not by creating homes for settlers. The federal lands now absorbed ex-servicemen in the capacity of Government Service employees of the agencies administering those lands.

The Multiple-Use, Sustained-Yield Act was followed in 1964 by the Wilderness Act.[26] Again, agency budgets, salaries, and careers benefited by the new law and the basis was laid for new regulation that would be used to infringe or confiscate the rights of stockmen on the federal lands—specifically his water.

In 1971 Congress passed the Wild Horses and Burros Protection Act purportedly to protect wild free-roaming horses and burros on the federal lands from abuse and destruction.[27] Like the Multiple-Use, Sustained-Yield Act and the Wilderness Act, the original intent and stated objectives of the Wild Horse Act were quickly exceeded. By any standards the mismanagement of wild horses and burros under the Act has become its hallmark. However, the Act has served to expand budgets, salaries, and careers of both the Forest Service and BLM and infringe and devalue the property rights of ranchers on the federal lands.

The Federal Land Policy and Management Act (FLPMA) was passed in 1976.[28] FLPMA was basically a creation of the Interior Department. It was drafted by the department with direct influence from the new conservationists.

THE ENVIRONMENTALISTS EMERGE

The conservation movement was reborn after World War II and emerged in the 1960s as the environmental movement. Prominent among the various environmental groups was the Sierra Club which John Muir had created in the 1890s to fight the stockmen's preemptive right claims to the new forest reserves. Muir and the Sierra Club faded after the reserves were firmly established and the Sierra Club's conservation rhetoric no longer needed to justify the lock-up of the

West's resource base. Muir and the Sierra Club wrecked on the rocky shoals of power politics when their objection to a water reservoir site in Hetch Hetchy Valley of the Sierra was overruled by financial and industrial interests under the leadership of Gifford Pinchot.

For many years—from the 1920s to the 1950s—nature preservationism languished with little public attention. Old line groups such as the Audubon Society (founded 1905) were joined by small and quiet organizations such as the National Parks and Conservation Association (1919), the Izaak Walton League (1922), the Wilderness Society (1935), and the National Wildlife Federation (1936). Their efforts received scant attention in the Depression era or during World War II. However, during the mid-1950s the movement began to awaken. The interested reader should examine Ron Arnold's reassessment of the environmental movement, *Ecology Wars*, for an in-depth analysis of relevant events.[29]

By the 1960s the older conservation movement's nature preservation branch began to evolve into the modern environmental movement. The change was more than in name. The newer movement embodied altogether different beliefs.

Another branch of the environmental movement that emerged in the 1960s was the Natural Resources Defense Council (NRDC). NRDC was the primary litigating arm of the new environmental movement. Sierra Club and NRDC, along with a host of other environmental groups, rallied around the creation of the Federal Land Policy and Management Act (FLPMA). Through FLPMA they hoped to accomplish the long term objectives of the northern core interests who financed their efforts: the elimination of private rights in the federal lands.

The political agenda of the environmental movement has always been oriented toward the promotion of centralized government and the limitation of private property rights. In FLPMA many environmentalists thought they had accomplished this objective. Environmentalists began to encourage the BLM and Forest Service to take an aggressive stance against the rights of ranchers. Chief among their targets was water arising on the federal lands and the facilities associated with them, in other words, the prior appropriation doctrine of the western states, specifically irrigation and stock water rights.

Behind their strategy was a methodology first developed by the Forest Service after 1916. That strategy, simply stated, was to use the vague and ambiguous mandate of the Act of June 4, 1897, "to protect the preserves," as a basis to demand livestock reductions or costly management practices that served to increase the rancher's cost-of-use. Theoretically, if the units of production, (cattle, sheep) are reduced while the cost per-unit of production is increased, a point will be reached where it costs more to graze livestock on the federal lands than the cow or sheep can return to its owner in profit. If the owner of livestock is faced with an ongoing loss in his federal land grazing operations he will eventually be forced to abandon grazing on those lands. He would also be abandoning his stock water rights and any other property rights associated with them. Since the ownership of water is a *usufruct* right (a right which depends on its continual beneficial use) the Forest Service could force abandonment of the rancher's preemptive right claims.

This approach to the destruction of the rancher's preemptive right claims had distinct advantages. Even though the rancher complained of economic damage and the *taking* of property, the federal agency could hide behind the *resource damage* argument and their congressional mandate to *protect* the resource. Since the federal agency controlled all the data and information about the resource, the stockmen were left with little recourse except to acquiesce to agency actions.

TAKINGS

The Fifth Amendment of the U.S. Constitution says: "No Person shall be . . . deprived of . . . property, without due process of law; nor shall private property be taken for public use without just compensation."[30]

The Forest Service developed the strategy of taking the rancher's property while hiding behind the resource protection mandate of the 1897 law which outlined their administrative duties. The Forest Service maintained it was not taking property, only protecting the resource. The permit, they said, was granted by the federal government and was revokable at the will of the government. Therefore, the permit was not property. As for the pre-existing property rights that were the basis upon which the original permits were granted, the Forest Service was silent. They maintained the premise upon which the reserves were created. The reserves had erased all those prior rights, they claimed, despite the U.S. Supreme Court ruling.

The Forest Service offered an administrative appeals procedure that was designed to handle complaints against the agency. In the administrative appeals process, the Forest Service sat not only as arresting officer, but also as judge and jury. The stockman found himself totally at the mercy of the Forest Service. Without substantial political intervention the stockman could find himself embroiled in an almost endless procedure to "exhaust the administrative remedies." This could take years and be extremely expensive. Without exhausting his administrative remedies he could never get his case into the federal courts where he could be granted the constitutional guarantee of due process.

The administrative appeal of the Forest Service allowed no evidentiary hearing. In other words, the Forest Service could bring in any number of witness against the stockman, but the stockman could not cross examine them. The record of the hearing was controlled by the Forest Service. There was no independent third party to weigh the dispute. The philosophy of agency self-protection pervaded the entire hearing process.

If the stockman could effectively raise a true *taking* or *economic damage* issue he could by-pass the Forest Service administrative appeals and go directly to the federal courts. Forest Service strategy has been to contain those complaints by settling disputes before they could go to a court, even if it meant substantial concessions to the plaintiff. By allowing only *resource damage* issues to proceed to the federal courts the USFS has been able to build an impressive bulwark of case law in their favor. The underlying pre-existing rights denied the constitutional guarantee of due process of law to the stockman.

With the passage of the Federal Land Policy and Management Act, October 21, 1976 (FLPMA), each agency engaged in a frontal assault on the rancher's stock water and irrigation waters arising on the federal lands. The Forest Service and BLM began to file claims on private waters, springs, seeps, creeks, wells, reservoirs, etc. They demanded special use permits for the rancher's use of ditches, pipelines, and flumes on the federal lands that had become private property recognized by Congress through the Act of July 26, 1866. Agency personnel began to demand quiet surrender of the rancher's rights, proclaiming themselves the "Lords of the Land."

The euphoria among the agency personnel and the environmentalists was short-lived, however. FLPMA contained one saving provision: "All actions by the Secretary concerned under this Act shall be subject to valid existing rights."[31]

In an unpublished opinion, the Tenth Circuit Court upheld a Colorado District Court ruling on November 7, 1977, affirming the validity of property rights acquired under the prior appropriative right doctrine and affirming the validity of the Act of July 26, 1866.

RIGHTS OF WAY

The Act of July 26, 1866 stated that once a person acquired water rights in accordance with local law and custom he may then secure a right of way across federal land for constructing ditches, flumes, canals, or pipelines to use the water. This act gave broad discretion to the right of way owner to build and maintain his water conveyance system. It recognized preexisting rights based on local laws, customs and court decisions. It did not even set boundaries for the right of way but allowed the owner to use that portion of the federal land and the materials thereon he felt were necessary to accomplish the purpose for which the right of way had been granted. The only restriction placed on a right of way owner was that he compensate for any damage his activities might cause to another settler. The act also provided that the control of water on federal lands would be under state governments, not the federal government.

Subsequently, the western states have developed vast bodies of water law which convey water rights to individuals. Water rights are private property and are treated by the law as real property as long as the water is put to the beneficial use for which the water right was granted. Beneficial uses include municipal, quasi-municipal, domestic, irrigation and stock watering. The beneficial use usually takes place when the water is transported from federal land to private land, but in some cases such as mining, hydropower production and stock watering, the beneficial use may take place in whole or in part on federal lands. To bring water to the place of its beneficial use, owners must often cross federal lands by ditch, pipeline or flume through these rights of way, granted originally by the Department of the Interior under terms of the 1866 act.

These rights of way are crucial to the survival of the West. They are not confined to water conveyance. Section 8 of the 1866 act provides for rights of way for roads and highways. Virtually all of the West's road and highway systems have their origin in Department of the Interior right of way grants. Virtually all of the electric power lines, telephone lines and underground cables depend on these rights of way. Natural gas pipelines, bulk fuel lines and coal slurry pipelines are largely dependent on old Interior Department right of way grants. These old grants continue to provide the legal basis for the existence of the West's transportation and communication system across federal lands. These right of way grants allow state or county road maintenance crews or private contractors to perform construction or maintenance within these rights of way without fear of interference from a federal agent or agency. The right of way owner can use his own discretion without obtaining permission from the federal government. If this were not so, even emergency repairs could be held up while a federal agency responded to local requests. This is precisely the problem Congress was trying to avoid when it passed the other acts providing for rights of way.

An 1891 right of way law passed by Congress required subsequent grants to have clearly defined marginal limits. It specified a strip of land extending fifty feet from the marginal limits of the ditch on either side of the ditch. Some settlers had taken advantage of the original lack of specific boundaries as an excuse to remove valuable timber or other material from federal lands for purposes unrelated to the 1866 act. However, the 1891 act provided that material might be taken from government land outside the right of way corridor if it was necessary for the construction of the ditch, flume or pipeline involved.

Congress passed another law February 15, 1901 to facilitate demands of major irrigation projects under the reclamation acts. It also clearly defined a right of way of fifty feet on either side of the water conveyance structure.

Grants made under the 1866 act, the 1891 act and the 1901 act were permanent. They are considered conditional fee interests in the land itself. The only interest retained by the U.S. in the rights of way are reversionary interests. If the right of way owner permanently abandons the ditch, flume or pipeline, the right of way cannot be used for another purpose, but reverts back to the United States.

Some had argued that the 1891 act creating the forest reserves had repealed the 1866 act because it made no mention of water rights. The 1901 law was also said to have repealed the water right provisions of the 1866 law because it too failed to mention water rights.

The Tenth Circuit Court ruled in 1977 that no subsequent act superseded or repealed the 1866 act, and that it was possible for anyone having a vested water right under local law or custom to secure a right of way for conveyance of that water under the 1866 Act.[32]

The actual language of the 1866 act that set forth the criteria for governing all of these rights of way was codified in the Federal Revised Statutes as R.S. 2477. To carry out the will of Congress, the Department of the Interior promulgated regulations under R.S. 2477 in Bureau of Land Management Instruction Memorandum 90-589 of August 15, 1990.

When Congress passed FLPMA in 1976, it repealed the right of way provision of the 1866, 1891 and 1901 laws for any new grants. However, it did not repeal existing grants. Confusion arose over which agency was to administer the old rights of way and under what rules. The major conflict was between the Interior Department and the Forest Service. The Forest Service tried to apply its own regulations to Interior Department rights of way. These regulations were highly restrictive. They demanded that a permit be acquired from the Forest Service and that a special fee be paid. They demanded that any maintenance on a right of way be done according to their specific requirements for the job in question. The Forest Service would have to be notified and approve any project before work could begin.

Where citizens had used their own discretion in moving or removing earth, stone or trees and other vegetation from their rights of way for over a hundred years, they were now being second-guessed by the Forest Service. Fines were imposed for violating directives of federal agents. Criminal actions were threatened by the Forest

Service.

Many rights of way were vested long before the creation of national forests. These rights of way were permanent withdrawals from the federal lands. Even a patent conveying property from the federal government could not override an existing 1866 right of way. All patents conveyed by the federal government after July 26, 1866 carried the exception language, "Excepting therefrom all rights of way for ditches or canals constructed by the authority of the United States." The acts creating various national forests contained language providing for the exclusion from the national forest of any previously withdrawn lands.

Right of way holders throughout the West sought clarification through the courts and legislative bodies. In 1982 the Tenth Circuit Court of Appeals ruled in *Denver v. Bergland* that the Forest Service could not impose rules on rights of way authorized by the Department of the Interior which would diminish the original grant. The Forest service had argued that the City of Denver had departed from their original right of way through the forest while converting their water conveyance system from ditch to pipeline. The court ruled that the Forest Service did not have the authority to interfere with a grant made by the Department of the Interior and that even if a departure was made from the original right of way, there was no violation as long as the purpose for which the right of way had been granted was being served.

In 1986 a bill sponsored by Senator William Armstrong of Colorado was passed as Public Law 99-545. It was intended as a clarification of the rules and regulations governing rights of way across federal lands. P.L. 99-545 expressly excluded the old Interior Department rights of way through the national forests from any Forest Service rule that might diminish the original grant.

After the 1986 law, the Forest Service developed regulations governing the rights of way it administered. Three Forest Service documents, Manual 2720, Manual 5500 and Handbook 5509.11 refer to the rights of way as permanent. They are not subject to post-FLPMA rules. Forest Service rules say the 1866 rights of way are to be managed according to R.S. 2477.

SUMMARY

It was just another in a long line of laws, regulations and court rulings spanning more than a hundred years that uphold the validity of private water rights on federal lands. Many of these water rights were stock water rights. The proof of beneficial use upon which these rights gained validity according to local law and custom and court decision was livestock grazing. Most of these valid rights predated the Forest Reserve Act of 1891 and the Taylor Grazing Act of 1934. Most of them predated the repeal of the preemption laws in 1891.

Other questions arose. If the stock water rights on the federal lands, established through livestock grazing as a proof of beneficial use, are valid, how can the right to graze be invalid? How can a valid right develop from an invalid premise? How can a valid right be maintained on an invalid premise? How can the stock water rights on federal lands be recognized as real property by the courts without recognizing the preexisting right to graze as a property right with equal substance?

Chapter Ten Footnotes

1. 48 Stat. 1269, "Act of June 28, 1934," codified at 43 U.S.C. 315 *et seq*. Commonly known as the Taylor Grazing Act. See Section I.

2. Ibid. Section III.

3. Ibid. Section IV.

4. *Kansas* v. *Colorado*, 206 U.S. 46 (1906).

5. 48 Stat. 1269, "Taylor Grazing Act," Section III.

6. *Broder* v. *Water Company*, 101 U.S. 274, at 276. (1879).

7. Testimony of Secretary of the Interior Harold L. Ickes, Taylor Grazing Act Senate Hearings, *Legislative History*, 1934.

8. 48 Stat. 1269, "Taylor Grazing Act," Section III.

9. Ibid.

10. *Mollohan* v. *Gray*, 413 F. 2d 349. (Decided June 17, 1969 by the Ninth Circuit Court of Appeals).

11. *Acton* v. *United States*, 401 F. 2d 896. (Decided October 4, 1968 by the Ninth Circuit Court of Appeals).

12. *United States* v. *Cox*, 190 F. 2d 293. (Decided June 21, 1951 by the Tenth Circuit Court of Appeals).

13. *Oman* v. *United States* 179 F. 2d 738. (Decided December 19, 1949 by the Tenth Circuit Court of Appeals).

14. *Bowman* v. *Udall*, 243 F. Supp. 672 (1965).

15. *Placer County Water Agency* v. *Jonas*, 275 Cal. App. 2nd 691. 80 Cal. Rptr. 252 (1969).

16. *Fauske* v. *Dean*, 101 N.W. 2d 769. (1960).

17. *United States* v. *Fuller* 409 U.S. 488. (Decided January 16, 1973). The Supreme Court reversed 442 F. 2d 504. (Decided April 9, 1971 by the Ninth Circuit Court of Appeals).

18. *Faulkner* v. *Watt*, 661 F. 2d 809. (Decided November 19, 1981, by the Ninth Circuit Court of Appeals).

19. *Barton* v. *United States*, 609 F. 2d 977. (Decided November 20, 1979 by the Tenth Circuit Court of Appeals).

20. *Chournos* v. *United States*, 193 F. 2d 321. (Decided December 20, 1951 by the Tenth Circuit Court of Appeals).

21. *Red Canyon Sheep Co.* v. *Ickes*, 98 F. 2d 308. (Decided May 27, 1938 by the United States Court of Appeals for the District of Columbia).

22. Libecap, *Locking Up the Range: Federal Land Controls and Grazing*, pp. 36, 42-61 *passim*.

23. *U. S. Statutes at Large*, The Employment Act of 1946, 79th Cong., 2d Sess., Ch. 33, pp. 23-26.

24. Ibid.

25. 74 Stat. 215, Multiple-Use, Sustained-Yield Act of 1960, codified at 16 U.S.C. 528-531.

26. 78 Stat. 890, Wilderness Act of 1964, codified at 16 U.S.C. 1131-1136.

27. 85 Stat. 649, Wild Horses and Burros Protection Act, codified at 16 U.S.C. 1331-1340.

28. 90 Stat. 2743, Federal Land Policy and Management Act of 1976, codified at 43 U.S.C. 1701.

29. Ron Arnold, *Ecology Wars: Environmentalism as if People Mattered*, Free Enterprise Press, Bellevue, 1987.

30. *United States Constitution*, Fifth Amendment. The subject of takings will be discussed in detail in Chapter 12.

31. 90 Stat. 2743, Federal Land Policy and Management Act of 1976, codified at 43. U.S.C. 1701.

32. ——————— v. U.S., Federal District Court for Colorado, 14, January 1976. Upheld by Tenth Circuit Court of Appeals. Unpublished Opinion, Nov. 7, 1977. [Plaintiff's name may not be cited.] Referenced in Bureau of Land Management Bulletin 85-45.

PART FOUR

RETHINKING RANGELAND

11

ECONOMIC VANDALISM

In 1942 Congress amended the Taylor Grazing Act. The nation was at war and land for the expansion of training and defense purposes was needed. A good portion of this land was rangeland. It was split estate federal land where the rancher owned the water and the right to graze cattle, but land on which the federal government had never recognized private title.

The amendment:

> Whenever use for war or national defense purposes of the public domain or other property owned by or under the control of the United States prevents its use for grazing, persons holding grazing permits or licenses and persons whose grazing permits or licenses have been or will be canceled because of such use shall be paid out of the funds appropriated or allocated for such project such amounts as the head of the department or agency so using the lands shall determine to be fair and reasonable for the losses suffered by such persons as a result of the use of such lands for war or national defense purposes. Such payments shall be deemed payment in full for such losses. Nothing contained in this section shall be construed to create any liability not now existing against the United States.[1]

The amendment, which required payment for grazing rights taken by the Department of Defense, was tested and upheld in the case of *McDonald* v. *McDonald*. [2]

In 1968 the assessor of Modoc County, California, imposed a possessory interest tax on federal grazing allotments.[3] Stockmen in California opposed the tax on the grounds that a grazing permit was a revocable privilege granted by the federal government and did not constitute taxable property.

The California Court of Appeals ruled against the stockmen, holding in 1971 in *Board of Supervisors of County of Modoc* v. *Archer* that grazing permits amount to a "possessory interest" within the meaning of the statute providing that possessory interests in land or improvements are taxable.[4] And in 1976 in *Dressler* v. *County of Alpine* the Court of Appeals further held that United States Forest Service grazing rights which taxpayers held for a number of years did constitute taxable possessory interests.[5]

Thus, since 1971 the stockmen of California have, by regulatory code, paid a possessory interest tax on their federal grazing allotments under the jurisdiction of the Bureau of Land Management and the Forest Service.[6]

The Internal Revenue Service (IRS) considers grazing rights as part of a stockman's estate for tax purposes.[7]

The courts have consistently upheld the IRS in taxing the value of grazing rights on the federal lands as part of the rancher's real property.

A typical grant bargain and sale deed transferring a ranch with federal range contains the following language:

> The undersigned have agreed, and do hereby agree, to a sale by the Sellers to the Buyers of certain real and personal property, water rights and privileges, grazing leases, rights and privileges more particularly described below, for the total purchase price of . . . .

The value of the grazing rights are capitalized into the value of the ranch unit for sale or tax purposes. This same language defines the collateral when a ranch unit is financed by a lending institution.

The United States Court of Appeals for the District of Columbia has said in *Red Canyon Sheep Co.* v. *Ickes*:

> [T]hose qualifying under the Act definitely acquired ''rights'' . . . Yet, whether they be called rights, privileges, or bare licenses, or by whatever name, while they exist they are something of real value . . . which have their source in an enactment of the Congress.[8]

In *McNeil* v. *Seaton*, in a dispute over a rule promulgated by the Secretary of the Interior that rancher Wade McNeil alleged had deprived him of grazing rights, the United States Court of Appeals for the District of Columbia said:

> To the extent that the Special Rule is repugnant to the Act and to the Range Code prior to 1956, if and as applied *to this appellant* to deprive him of his rights and to award others grazing privileges in derogation of appellant's preference status, it is void.[9] [Italics in the original]

In *Osborn* v. *United States* the Ninth Circuit Court of Appeals held that a regulation that a grazing preference is not a property right does not mean that a permit is not a property right and can not be so construed.[10]

In *Hubbard* v. *Brown* the California Supreme Court in 1990 rejected an argument that federal regulations deny any interest in federal lands to grazing permit holders.[11]

Given the preponderance of evidence that grazing rights are property, that grazing rights have value, that grazing rights are owned by the rancher and not the federal government, how have the BLM and Forest Service managed to avoid the *taking issue*? How have they been able to deny the rancher *due process* and *just compensation* as guaranteed under the Fifth Amendment of the U.S. Constitution? How have they been able to deny him equal protection under the law as guaranteed by the Fourteenth Amendment of the Constitution?

The resource damage argument that was discussed in Chapter Nine is part of the answer, but that alone does not explain the success of the BLM and Forest Service in avoiding paying compensation or an adverse judgement for taking valuable property.

THE PROCESS OF TAKING

When a taking of a rancher's range rights is effected, it usually involves all or some of the following elements:

1. The taking is premised on the resource damage or resource protection argument.

2. The taking is justified as necessary for the public good.

3. The taking is accomplished by "agreement" with the stockman.

4. The taking action involves using such terms as *suspended*, *non-use*, *partial*, or *temporary*. These terms avoid the finality of a mandatory unilateral elimination of the grazing rights.

The taking of all or a portion of a rancher's range rights is usually preceded by a "study" and the accumulation of range "data" which will be used to support the resource damage or resource protection argument. The efforts to mitigate resource damage or to enhance resource protection are always reinforced by alluding to the public interest.

A unilateral reduction of grazing rights could raise the taking issue in the event the rancher was able to successfully challenge the resource argument and its *data* base. For that reason the agency usually tries to elicit an agreement from the rancher. The rancher understands the range value and its vital relationship to his ranch unit. He must defend it at all costs. On the other hand, the resource damage or protection issue has effectively denied him access to his Fifth Amendment right of due process. As long as the argument is confined to resources and a permit to graze, he is faced with substantial case law denying him compensation. The rancher finds himself in the position of signing an "agreement" with the agency for a reduction in his grazing rights in order to avoid an even greater reduction the agency threatens if he resists. Once he signs an agreement the taking issue is effectively dead.

If the rancher resists the efforts of the agency to reduce his grazing rights, the agency can issue a decision to reduce the grazing capacity of the range. Again, if the justification for the reduction has been addressed solely in resource terms, the argument will probably be confined to the administrative appeals process. If the rancher has failed to raise the counter issue of economic damage and lack of due process he probably has no effective recourse to a court of law even if he exhausts the administrative appeals process.

The agency, on the other hand, is able to argue that the reductions in grazing did not completely eliminate the rancher's grazing. Being only a partial reduction it has not "taken" any property, even if the pre-existing right argument has been raised by the rancher. In other words, even if the rancher is successful in establishing the fact that real property is involved, if the reduction is less than his full grazing right, it helps defuse any taking question. If the reduction is expressed in terms of suspended, temporary, partial, or non-use, the implied reinstatement of full grazing capacity further reduces taking implications, even if the reduction becomes permanent which it almost without exception does.

CASE LAW AND TAKING

Until 1876 there was some doubt as to the existence of a federal power of eminent domain.[12] In the case of *Kohl* v. *United States*, the issue was resolved. The United States Supreme Court held that the powers vested by the Constitution in the general government demanded for their exercise the acquisition of lands in all the States. Further, no State could enlarge or diminish that power.[13]

Due process of the law which is inherent in the taking argument must involve compensation: "a form of procedure to be observed in the taking of private property for public use . . . is not due process of law if provision be not made for compensation."[14]

It is a violation of the Fourteenth Amendment and a deprivation of due process if property is taken for other than a public use.[15]

The Fifth Amendment guarantee "that private property shall not be taken for public use without just compensation was designed to bar government from forcing some people alone to bear public burdens which, in all fairness and justice should be borne by the public as a whole."[16]

The just compensation required by the Constitution is, "A full and perfect equivalent for the property taken."[17] The general standard thus is the fair market value, what a willing buyer would pay to a willing seller.[18]

If a unit of real property is condemned, the market value of that property must be paid to the owner. The franchise of a private corporation is property which cannot be taken for public use without compensation.[19] Letters patent for a new invention or discovery in the arts confer upon the patentee an exclusive property for which compensation must be made when the government used the patent.[20]

Government requisitioning of electric power produced by water thereby cutting off the supply of a lessee entitled the lessee to compensation for the rights taken.[21] Where regulatory action requires the cessation of an activity with or on the property or deprives it otherwise of value, whether there has been a taking or not in the Fifth Amendment sense becomes critical. But the government decided long ago that taking included destruction or severe impairment of use.[22]

The anti-livestock grazing advocates have supported partial taking and temporary taking through agency regulatory activities to infringe or destroy the value of the rancher's range rights since early in the twentieth century. Regulatory taking of private property has been an underlying *de facto* element associated with federal management of the range.

Let's briefly analyze some of the actions of the BLM and Forest Service against ranchers to see if they in fact involve a taking of property. Let's first view this from a purely practical point of view. Does the rancher in fact suffer economic loss as a result of government action?

GRAZING FEES AND TAKING

Consider the grazing fee controversy and taking.

This controversy has its roots in the first efforts to impose a tax on livestock grazing in the forest reserves. Almost from its inception there have been ongoing efforts by the government and anti-grazing interests to increase the fee.

Pricing the stockman off the range has been the clear objective of many anti-grazing advocates over the years.

The argument is usually broached in comparison with private lease rates. Private pasture, it is argued, rents for four or five times the amount the stockman pays the federal government in a grazing fee. Oftentimes the federal range is very close in quality to private range. In some cases it is superior. The proponents of an increased grazing fee contend that the rancher should be paying the federal government a fee comparable to what he would have to pay a private landowner for a lease on private pasture. The private pasture rate establishes fair market value for grazing, therefore if the rancher pays less than the private lease rate, he is not paying fair market value. His lower grazing fee, they contend, amounts to a subsidy.

The difference between the lease rate for private pasture and the grazing fee on federal lands involves the question of who is providing what inputs and the value of those inputs. A grazing operation, whether it be situated on federal land or private land, involves the same basic inputs. Forage, water, water facilities, livestock handling facilities, supplemental feed, accommodations for personnel, and management are among the major inputs to any livestock grazing operation whether it be on private land or federal land. The fair market value of any lease depends upon how many of the necessary inputs are provided by the lessor and how many by the lessee, and the comparative value of those inputs.

In a typical private pasture lease, all of the above mentioned inputs are provided by the landowner or lessor. The stockman transports his livestock to the lessor's premises. The lessor accepts responsibility for the livestock while they are on his premises, sometimes to the extent of being responsible for death loss and the rate of gain. The lessee has exclusive use of the lessor's pasture for the term of the lease. His primary responsibility is to transport the livestock to the premises at the beginning of the lease term and to remove them from the premises at the termination of the lease period.

For all these services and the exclusive use of the lessor's pasture he pays a lease rate which reflects the cost of the inputs being provided by the lessor. Obviously, if the lessor is providing all or most of the cost of the livestock grazing operation, the lessee is relieved of these direct costs, having paid a proportionate amount for the lease.

A grazing operation on federal land competing with other federal land users involves the same inputs and costs that accrue to a grazing operation on private land. But in this case the inputs are almost totally provided by the rancher. The federal government as landlord provides virtually none of the basic physical inputs or operating expenses.

The rancher may own the water rights on the federal lands, not the federal government. The water is an absolutely essential input. Without the water, the land has no value for grazing. The rancher provides the forage by virtue of the fact that he owns a pre-existing grazing right. That was the proof of beneficial use that gave him the water under local law and custom and court decisions. Even if we look at the forage from the point of view of those who do not recognize the pre-existing right and view the grazing fee as a lease rate, we still must conclude that the rancher provides the forage. If his grazing fee is a lease rate rather than a form of a tax on his pre-existing rights, then he paid for the forage when he paid for the lease and therefore owns the forage for the term of the lease.

Water facilities are usually owned by the rancher, livestock handling facilities are usually owned by the rancher, supplemental feed is provided by the rancher, personnel accommodations are provided by the rancher as well as management. Many of the inputs necessary to provide an income stream from the federal land are the facilities which exist on the commensurable base properties which are private lands. It is a clear rule of appraisal that the cost of use of any property are those costs which contribute to the income stream of the property. Much of the value of federal land grazing derives from the facilities on commensurable private land. Much federal grazing land would have little or no value if it were not for the dependent base properties of the rancher.

It also follows that the small acreages of private commensurable base properties derive a considerable share of their value from the contribution made by the federal range which adjoins it. Any factor which tends to increase costs, or reduce income from the commensurable federal range, has the impact of reducing the value of the rancher's private commensurable fee simple base lands.

The value of any business is a function of capitalized net income. Any action of the BLM or Forest Service which reduces the income of a ranch, without a proportionate decrease in production costs, reduces the value of the entire ranch.

To arrive at fair market value of federal grazing land a gross value would have to be established by comparison with private lease rates. From that gross value must be deducted the value of the inputs provided by the rancher. The remainder represent fair market value of the lease.

For example, in a joint BLM/Forest Service appraisal of western range lands conducted in 1983, the Forest Service and BLM concluded that the gross value of central Nevada range was $4.68 per animal-unit-month (AUM). In an estate appraisal in the same area, to satisfy Internal Revenue Service (IRS) requirements, it was determined that the value of the rancher's rights in the federal lands amounted to $4.50 per AUM. The remaining eighteen cents then, would represent the government's share of the gross lease rate or fair market value to the federal government. This was substantially less than the grazing fee charged by the federal government.

The value of the inputs provided by the rancher on federal grazing land can be derived from Schedule F of the Federal Income Tax Return. The ranch operating costs, listed on Schedule F, will stand a court test. These are the legal inputs that must be cited in any legal confrontation over the cost of use on federal lands. By using court-sanctioned appraisal procedures to derive the fair market value of federal land grazing, the figure almost always is lower than the current grazing fee. In some cases, particularly where the agencies have been employing excess regulation, the fair market value of federal land grazing is a negative figure.

The subtle application of costly regulation to the rancher's grazing allotments have rendered many grazing allotments economically marginal. A small increase in the grazing fee can quickly render a grazing allotment completely uneconomic. Under these circumstances the value of the permit (see *Red Canyon Sheep Company* v. *Ickes*) has been destroyed.[23] The rancher has suffered a genuine economic loss and the taking issue immediately arises.

If the rancher no longer has an economically viable grazing operation on the federal lands, he can no longer afford to utilize his stockwater on the federal lands. He is deprived of the income from his real property. Water rights, being usufruct rights, demand continued use for the right to be maintained. If the rancher is deprived of the economically viable use of his water for a sufficient period of time he can lose that real property. From a practical point of view then, an increase in the grazing fee can result in the taking of real property.

COMMENSURABILITY AND TAKING

The rancher's base property, his fee simple commensurable land base, is dependent property. Commensurable private lands were recognized by the Forest Service as dependent on the federal range for a major portion of their value. This factor was instrumental in the dispute over grazing that almost forced the repeal of the forest reserves in 1897. This dependent private commensurable land issue was at the heart and core of the rules upon which the first grazing permits were granted.[24]

The Taylor Grazing Act also recognized that commensurable private lands were dependent on the federal range.

A significant portion of the income stream of the base property is derived from the commensurable federal range. If this were not so, there never would have developed a range livestock industry in the western United States. Another way of stating it is that if the base properties had been economically viable units without range and range water rights on the federal domain, then there would have been little incentive to extend the grazing operation beyond the private base property boundaries.

Any uncompensated increases in costs on the federal range, such as increased grazing fees, have the direct impact of reducing the value of the commensurable dependent base property. Certainly it constitutes an economic loss to the rancher. Whether or not it is a taking, in the legal sense, could be reflected in the degree of value lost.

THE COMPETITIVE BID AND TAKING

Proposals to put the present grazing allotments of western ranchers up for competitive bid reflects a lack of understanding of the taking issue and the split estate nature of the federal rangelands. Any administrator of a competitive bid program would be forced to consider the contribution of the rancher's private property rights in the federal lands such as water and range improvements. A competitive bid would have to recognize the contribution of the base private property to the income stream of the federal rangelands. The gross value of the lease rate as determined by competitive bid would have to be split proportionately with the rancher who owns the pre-existing rights on the federal lands and the commensurable base property. In most cases, this would leave little or no return to the federal government from the lease. Both the successful bidder and the federal government would be forced to recognize that the federal range is not a separate independent entity, but an integral part of a long-established mutually dependent two-component ranch unit.

If the rancher's interests in the range were ignored, this would clearly imply a taking under the Fifth Amendment of the Constitution--both for the private values in the federal lands and the reduced value of the dependent base property.

SUB-LEASE REGULATIONS AND TAKING

In recent years administrators have attempted to prevent sub-leasing of federal grazing allotments to a second party. The Forest Service has long had rules limiting sub-leasing. The BLM also requires that any sub-lease revenues from a grazing allotment in excess of the current grazing fee be returned to the government. But what is the rancher really leasing in a federal grazing land sub-lease?

The rangelands are split estate lands. The rancher is entitled to a return on, and a return of, his investment in the water rights and range improvements. He is entitled to fair compensation for any management he contributes. If private commensurable base property is involved he is entitled to a fair return from that contribution. An attempt to prevent a fair return to the rancher from a sub-leased grazing allotment is an obvious attempt to deprive the rancher of his legitimate property rights. The rancher is clearly suffering economic loss from the agency action.

RESOURCE ISSUES AND TAKING

When wildlife species are introduced on a rancher's grazing allotment without his express consent and in competition with the livestock on his ranching operation, the question of increased costs and decreased revenues arises. The value of any business--and ranching is a business--is a function of capitalized net income. If the introduction of game species by the agencies effects the rancher by increasing costs, reducing income, or both; he has definitely suffered an economic loss at the hands of the federal agency.

When the riparian area protection argument is used to deny a rancher access to his water or the use of forage, it can constitute tangible economic damage. When wild horses or burros are allowed to encroach on the water and forage the ranch depends on, it can create economic damage. When an "endangered" species or "area of critical environmental concern" blocks a rancher from a portion of his range or water, it can decrease net income. A decrease in net income, if sustained and without compensation, is a devaluation of the ranch unit. It is a forced reduction in property values.

Do the anti-grazing advocates know this? When Johanna Wald, attorney representing the National Resource Defense Council (NRDC) in the 1986 *NRDC* v. *Hodel* grazing case was asked by the judge if she felt the grazing fee should be raised, she replied in the affirmative. When asked how much she felt the increase should be, Wald replied, "High enough to remove livestock from the range."

When Executive Order No. 12630 was signed by President Reagan on March 15, 1988—which prohibited regulatory taking of private property without compensation—it sent a wave of fear through the environmental movement. The Executive Order, they said, if not rescinded, threatened to destroy the entire environmental movement.[25]

"POLITICAL" RANGE MANAGEMENT

The taking of the rancher's split estate rights in the federal lands is usually accomplished by cloaking the action in "resource damage" or "resource protection" language. "Best available data" usually determines the outcome of agency administrative hearings. The agency can collect persuasive amounts of "data" at taxpayer expense. The rancher is denied this advantage. Consequently the stockman on the federal lands has been the loser in the vast majority of resource damage arguments.

An assertion by the federal agency that a range is overgrazed will be supported by an impressive "data" base. The rancher may know that the range is in excellent condition and that the overgrazing allegation is motivated politically, but without a data base comparable to that of the agencies, he has little chance of winning. An employee of the BLM or Forest Service with a degree in range management will probably qualify as a expert in any hearing even though his tenure in the area may be a matter of weeks or months. The rancher on the other hand will probably not be considered an expert on the range even though his formal education is superior to the agency employee and even though his tenure in the area spans a lifetime.

The use of "scientific range management" as a political tool began soon after the Forest Service emerged as an agency. By being in total control of the gathering, interpretation, and dissemination of range data, scientific range management was quickly adapted to the achieving of political ends.

"Scientific range management" by its very nature lends itself to abuse and manipulation because it is not a science in the rigorous sense. Scientific terminology is substituted for true scientific discipline. The rigid control of the scientific method, as would be employed in a true science such as chemistry or physics, seldom emerges. "Ocular reconnaissance" sounds scientific, but really means "visual observation," something the rancher does on a constant basis. "Use pattern mapping" produces impressive maps and overlays, but can be done from the desk in a Forest Service or BLM office without any observation on the ground itself. The same is true of "utilization studies" and "trend studies." As long as the agency has total control of the gathering and interpretation of the data, scientific range management can be manipulated to achieve predetermined political objectives.

The Forest Service openly endorsed political use of range data at a Ranger's Training Camp, May 7, 1956. Officers told trainees that neither moral or scientific correctness of a program was the determining factor in its implementation. The key to proceeding with programs would be "public acceptance."[26]

Chapter Eleven Footnotes

1. *U. S. Statutes at Large*, Ch. 500, 77th Cong., Sess. 2, July 9, 1942, pp. 654-655.
2. *McDonald* v. *McDonald*. 302 P. 2d 726. (Decided October 18, 1956 by the Supreme Court of New Mexico).
3. The possessory interest tax is codified in West's Annotated California Codes: Revenue and Taxation, § 107. California's basic possessory interest tax was enacted in 1939: see Cal. Stats. 1939, c. 154., p. 1277. § 107. The code states:

   "Possessory interests" means the following:
   (a) Possession of, claim to, or right to the possession of land or improvements, except when coupled with ownership of the land or improvements in the same person.
   (b) Taxable improvements on tax-exempt land.

   This statutory definition clearly acknowledges the split estate condition of the federal lands, which are tax-exempt. The grazing allotment tax was imposed by the California Board of Equalization as a regulation. See footnote 6 below.
4. *Board of Supervisors of County of Modoc* v. *Archer*, 18 Cal. App. 3d 717. 96 Cal. Rptr. 379. (1971).
5. *Dressler* v. *County of Alpine*, 64 Cal. App. 3d 557. 134 Cal. Rptr. 554. (1976).
6. Cal. Code Regs., Tit. 18, Reg. 28, "The following are examples of commonly encountered taxable possessory interests: ... (d) The right to graze livestock or raise forage on public lands." This regulation was codified in 1971 as a result of the *Archer* case, and amended in 1976 to reflect the *Dressler* decision.
7. *Shufflebarger* v. *Commissioner of Internal Revenue* 24 TC 980, 1955.
8. *Red Canyon Sheep Co* v. *Ickes*. 98 F. 2d 308 at 315. (Decided May 27, 1938).
9. *McNeil* v. *Seaton*. 281 F. 2d 931. (Decided June 16, 1960).
10. *Osborn* v. *United States*, 145 F. 2d 892. (Decided Nov. 24, 1944).
11. *Hubbard* v. *Brown*, 785 P.2d, 1183 (1990).
12. The taking issue is exceedingly complex. Useful sources are Ellen Frankel Paul, *Property Rights and Eminent Domain*, Transaction Books, New Brunswick, 1987; Richard A. Epstein, *Takings: Private Property and the Power of Eminent Domain*, Harvard University Press, Cambridge, 1985; Fred Bosselman, David Callies and John Banta, *The Taking Issue: An Analysis of the Constitutional Limits of Land Use Control*, Council on Environmental Quality, Washington, D. C., 1973.
13. *Kohl* v. *United States*. 91 U.S. 367, at 371. (1872).
14. *Chicago, Burlington & Quincy R.R. Co.* v. *Chicago*. 166 U.S. 26. (1896).
15. *Fallbrook Irrigation District* v. *Bradley*. 164 U.S. 112, 158-159. (1896).
16. *Armstrong* v. *United States*. 364 U.S. 40, at 49. (1960).
17. *Monongahela Navigation Co.* v. *United States*. 148 U.S. 312, at 326. (1893).
18. *United States* v. *Miller*. 317 U.S. 369, at 374. (1943).
19. *Monongahela Navigation Co.* v. *United States*. 148 U.S. 312, at 345. (1893).
20. *James* v. *Campbell*. 104 U.S. 356, 358.
21. *International Paper Co.* v. *United States*. 282 U.S. 399.
22. *Pumpelly* v. *Green Bay Co.* 80 U.S. 166. (1871).
23. *Red Canyon Sheep Company* v. *Ickes*. 98 F. 2d 308 at 315. (Decided May 27, 1938).
24. *Chournos* v. *United States*. 193 F. 2d 321. (Decided December 20, 1951 by the Tenth Circuit Court of Appeals).
25. *Public Lands News*, Summer, 1988.
26. 1956 Ranger's Training Camp, "Range Administrators," May 7, 1956, Box 3, 13/2/1:3, Record Group 95, Federal Records Center for the National Archives, Denver, Colo.

12

THE TAKING

Have the federal land management agencies and their environmentalist supporters adopted the intentional systematic erosion of the range livestock operator's property values as *de facto* policy? Have the loudly voiced goals of the Sierra Club, the Natural Resources Defense Council, Earth First!, and Free Our Public Lands! to eliminate livestock grazing from the federal lands been knowingly and intentionally translated into action by the federal land management agencies? If so, were they acting in violation of property rights of the stockmen?

The standard reply of the agencies when responding to the question of a taking of property under the Fifth Amendment as it relates to a rancher's grazing rights is: "A grazing permit represents a revocable privilege granted by the government and is not compensable as real property."[1] It is with this explanation that they decline to analyze their actions in light of the recent U. S. Supreme Court cases, *First English Evangelical Lutheran Church of Glendale* v. *County of Los Angeles,*[2] *Nollan* v. *California Coastal Commission,*[3] or Executive Order No. 12630 which instructs federal agencies to analyze their activities involving private property from the taking perspective.[4]

But such replies are usually given in the context of the *revocable permit*, and not the *underlying pre-existing rights* upon which the permit was granted. Such replies also generally ignore impacts on the value of the dependent commensurable private land base.

EXECUTIVE ORDER NO. 12630

On March 15, 1988, President Ronald Reagan signed an executive order titled *Governmental Actions and Interference With Constitutionally Protected Property Rights.* In four pages President Reagan set forth a general policy designed to protect private property and specific agency actions designed to enforce that policy. The executive order is not justiciable in itself, but does establish policy for the federal agency to follow. Additionally, it sets forth guidelines on issues of taking that *are* justiciable.

The Attorney General of the United States then issued to the Department of Justice staff a document titled *Attorney General's Guidelines for the Evaluation of the Risk and Avoidance of Unanticipated Takings* designed to enforce the president's order. A brief review of those guidelines is instructive.

Under definitions we find that "Private Property" includes "all property protected by the Fifth Amendment to the United States Constitution, including, but not limited to, real and personal property and tangible and intangible property."

Water rights are tangible real property. A grazing license or permit is personal property. Base property is tangible real property.

"Taking Implication" is defined as "Any policy or action to which the Executive Order applies that, upon examination by the decisionmaker . . . appears to have an effect on private property sufficiently severe as to effectively deny economically viable use of any distinct legally protected property interest to its owner, or to have the effect of, or result in, a permanent or temporary physical occupation, invasion, or deprivation, shall be deemed to have a taking implication for the purposes of the Executive Order and these Guidelines."[5]

Water rights, range rights, range improvements, and base properties are all legally protected property interests. An increase in the grazing fee can render all or part of these property interests economically non-viable. The placing of a rancher's grazing allotment in a competitive bid situation has the same effect, as does the denial of the right to sub-lease a grazing allotment at fair market value. Introduction of a game species onto a rancher's range can economically damage or destroy a ranch operation. Intrusions by uncontrolled wild horses and burros can create economic damage. The fencing of "riparian" areas or the exclusion of grazing to "protect the public interest" in "areas of critical environmental concern" or "endangered species habitat" can also create economic damage.

The *public good* argument does not justify a taking without compensation.

> The Fifth Amendment guarantee: "that private property shall not be taken for a public use without just compensation" was designed to bar Government from forcing some people alone to bear public burdens which in all fairness and justice, should be borne by the public as a whole.[6]

The Guidelines contain a section titled "Underlying Premises of the Fifth Amendment," which states,

> The Fifth Amendment provides that "private property [shall not] be taken for public use, without just compensation." Ownership, use, and transfer of private property of all types are rights. They are not benefits or privileges bestowed by government. At the same time, government also has the obligation to lawfully govern. Thus, the rights of property owners are not absolute and government may, within limits, regulate the use of property. Where those regulations amount to a taking of private property, government must pay the owner just compensation for the property rights abridged. The fact that the government's actions are otherwise constitutionally authorized does not mean that those actions cannot effect a taking. On the other hand, government may not take property except for a public purpose within its constitutional authority, and only then, on the payment of just compensation.[7]

ARE STOCK WATER RIGHTS PROPERTY?

The stock water rights on federal lands were created under the common law principle of first use and prior appropriation. The prior appropriation doctrine became law in the western states and Congress acknowledged the validity of that law. Is there case law on the topic of stock water rights as real property recognized by the federal government? In *Hunter* v. *United States*, the editors of the *Federal Reporter, Second Series* summarized the Ninth Circuit Court's opinion in keyed excerpts:

> Right to water by prior appropriation from public domain for any beneficial use is entitled to protection.[8]

But was livestock grazing a beneficial use for appropriation of water on the public domain? The *Hunter* court held that:

> Evidence by ownership of livestock that he and members of his family had since prior to 1880 persistently grazed and watered livestock in public domain . . . clearly showed a legal basis for acquisition of an appropriation to water by virtue of local decision.[9]

Were these stock water rights associated only with creeks flowing on to the private land where the grazing occurred or was grazing as a proof of beneficial use taking place on federal lands to secure legal rights to waters on federal lands? The Ninth Circuit in *Hunter* held that:

> In view of practically uncontroverted proof developed by individual establishing that he and his predecessors had appropriated water from public domain . . . and that they had always used all water that was available from springs and creek whose flow fluctuated from year to year, judgement protecting grant of use of waters should be shaped accordingly.[10]

The federal agencies have by policy denied Fifth Amendment and Fourteenth Amendment consideration of a rancher's range rights on the federal lands by asserting range rights are not "property" entitled to compensation. The question again emerges in the Ninth Circuit Court opinion in *Hunter* v. *United States*. If the stock water rights on the federal lands are held to be valid, how can the proof of beneficial use—livestock grazing—be invalid? If the pre-existing water right is valid, how can the grazing right upon which it depends for its validity not also be a valid pre-existing right? This is the same general question which emerged in the Supreme Court's *Buford* v. *Houtz* opinion. The Supreme Court's answer in 1890 preceded the desperate attempt to erase those preemptive rights through the establishment of forest reserves in 1891.* Forest Service and Interior Department policy have sought to evade that answer since March 3, 1891.

The implications of an affirmative answer to the range right question are enormous. No doubt those implications influenced the Ninth Circuit Court in *Hunter* v. *United States* to conclude that, even though Hunter's stock water rights were valid, the right to graze the surrounding lands was not essential to the use of the right.

*Supra, p. 97ff.

The decision in *Hunter* v. *United States* seems to clearly settle the pre-existing range right argument in the negative. Stockman Roy Hunter had argued that he had a right to graze cattle. He asserted his grazing rights were established by prior appropriation according to local law and custom and court decisions which had been reinforced by the Act of July 26, 1866, and *Buford* v. *Houtz*. The lands and waters in question had been withdrawn for a national monument. Hunter's grazing had been excluded. Hunter argued that the land surrounding his water was essential for the use and utilization of the water.

The court agreed that Hunter's water rights based on grazing use were valid, but also concluded that his right to graze lands adjacent to the water was not essential to the use of the water. The court reasoned it was possible for Hunter to continue the use of his water by piping it out of the boundaries of the monument. Therefore the land directly adjacent to the water was not essential to Hunter's enjoyment and use of his property. The court did not say Hunter had no grazing right. Rather, they affirmed that right. They said that the exercise of that right in his case was not essential within the boundaries of the monument.

Hunter was given the option of piping the water out of the monument or abandoning his water rights.

MUST TITLE BE TAKEN?

Apologists for BLM and Forest Service policy toward the stockman emphasize that grazing fee increases, invasion by introduced game species or regulations "protecting" the resource, do not constitute an actual taking of title of any of the rancher's property. The Justice Department guidelines are not so sanguine:

> Government has historically used the formal exercise of the power of eminent domain, which provides orderly processes for paying just compensation, to acquire private property for public use. However, government may become liable for the payment of just compensation to private property owners whose property permanently or temporarily has been either physically occupied or invaded by government or others with the assistance or approval of government, or so affected by governmental regulation as to have been effectively taken despite the fact that the government has neither physically invaded, confiscated, or occupied the property nor taken legal title to the property.[11]

The guidelines plainly acknowledge the force of the prior appropriation doctrine of the states that created the rancher's split estate interests in the federal lands.

> The Fifth Amendment's protection extends to all forms of property—real and personal, tangible and intangible. Property is not defined by the Constitution, but by independent sources such as state, local and federal law.[12]

The opinion in *Hunter* v. *United States* raises questions: if a stock water right is isolated in such a way that its value is lost by removing the water to a secondary site, is that a taking under the Fifth Amendment of the Constitution? If abandonment of the water is forced by federal rules or regulation, and its value destroyed, is this a compensable loss?

The court agreed with Hunter's argument, holding that the well-settled rule that the grant of a right in real property includes all incidentals possessed by the grantee and without which the property granted cannot be fully enjoyed.

The court disagreed with Hunter, holding that the adjoining lands provide the means to use the water beneficially and must therefore be deemed appurtenant to it. The court held further that the appurtenance must be limited to that which is *essential* to the use of the right granted.

The key word here is "essential." Hunter was allowed an easement or right of way to pipe the water out of the national monument where grazing could continue on other federal lands.

It is instructive to refer to water law authority Wells Hutchins regarding the broad topic of stock water on the federal lands and specifically how it relates to the place of use:

> Stockwatering is expressly listed in some of the water rights statutes as a purpose for which water may be appropriated. Courts have specifically recognized it as a beneficial use.
>
> Special attention is paid to stockwatering in some of the water rights statutes. A Nevada statute enacted in 1925 supplements the general water rights statute by prescribing certain conditions with respect to the acquisition of rights for watering livestock, particularly range livestock.[13] New Mexico extends to travelers the right to take water for their own use, and for their animals, from waters flowing from natural sources and also exempts from the statutory requirements for appropriating water those who may construct tanks or ponds having capacity of 10 acre-feet or less for the purpose of watering stock.[14] Oregon accords special treatment in case of an application to appropriate water for stock ponds of other small reservoirs from which water is not to be diverted or required to flow through the ponds.[15] South Dakota makes special provision for appropriations from minor streams for irrigation or stock purposes.[16]
>
> A Texas statute authorizes any landowner to construct on his own property a reservoir to impound not more than 200 acre-feet of water for domestic and livestock purposes "without the necessity of securing a permit therefor," but apparently requires him to obtain a permit if he desires to withdraw water from such impoundment for purposes other than domestic or livestock use.[17]
>
> The Nevada Supreme Court held that the general rule that to constitute a valid appropriation of streamflow there must be an actual diversion, does not apply to an appropriation for watering livestock in natural watering places formed by natural depressions.[18] The Nevada stockwatering act relates to particular watering places, to which the quantity of water

appropriated is measured by the number and kind of animals watered. It obviously contemplates use of the water in place, with no question about diverting it from a stream.[19]

The Hunter ruling was handed down in 1967 by the Ninth Circuit Court of Appeals. This was a period when Fifth Amendment guarantees were being loosely interpreted by our courts. This was a period when the Ninth Circuit Court of Appeals was upholding the taking of private property through regulation by the California Coastal Commission and various city, county, and state governments. In 1986 the U.S. Supreme Court laid the basis for the reversal of many Ninth Circuit Court rulings, with its holdings in *First English Evangelical Lutheran Church of Glendale* v. *County of Los Angeles* and *Nollan* v. *The California Coastal Commission*.

In light of these recent Supreme Court decisions on takings, It would appear that the opinion in Hunter may not withstand a well-constructed effort to overturn it.

If, as in the case of Hunter, the court would again avoid the taking issue by asserting the water can be transported to another site for grazing purposes and if the cost of transporting that water is economically unfeasible or physically unfeasible, the cost of transport being more than the value of the water for livestock, it would appear this would establish a taking. Not only would it establish a taking but by determining what constitutes a prohibitive cost of water transport or the prohibitive cost of changing the place of use, the value of the original grazing right has been established. Again we are forced to arrive at the inescapable conclusion: a pre-existing grazing right exists on these lands as private property. It does have value.

WILDERNESS AND WATER RIGHTS

Wilderness legislation is revealing of the awareness that exists among the members of the environmental community about the legitimacy of range rights and water rights as real property. Those who draft wilderness legislation sometimes make provisions for continued grazing activities on the proposed wilderness area and to disclaim any intent to affect water rights.

The concern about water and range rights expressed by the framers of wilderness legislation does not derive from the political strength of the livestock grazers or a benevolence on the part of environmentalists toward the livestock grazers. Wilderness legislation has been imposed on western states despite vehement protests by the states elected representatives and myriad user groups. Powerful user groups have been excluded from wilderness areas, groups who can claim a much broader political base than can the stockman.

A desire to support livestock interests from a benevolent perspective hardly explains environmental support of the rancher's grazing and water rights in wilderness areas. The people and groups promoting wilderness are often the same who carry on the anti-grazing propaganda in other quarters. Wilderness proponents are often the same people and groups who support *pricing* the rancher off the range through grazing fee increases. They are usually the ones supporting myriad other efforts to take range and water rights through regulatory overkill by the BLM and Forest Service.

Wilderness advocates clearly disclaim any intended adverse impact on grazing and water rights because an attempt to eliminate those rights through legislation would clearly be a taking and challengeable as such. They know water rights are property. They know the grazing permit clearly identifies the grazing area commensurable to the water as the site where the *use and enjoyment* of the water right takes place. Unlike the situation in *Hunter* v. *United States*, the cost of piping the water out of the wilderness area would be prohibitive, which in itself raises the taking issue. The necessary conduits and rights-of-way for the transport of water to another area could create disturbances negating the wilderness qualities the legislation purportedly protects. The privately owned water rights in some wilderness areas are extensive, amounting to dozens of springs, seeps, creeks, and underground sources. Where wilderness proposals have contained language protective of range and water rights, it has not been because wilderness proponents are necessarily friends of grazing interests. It is primarily because they want to avoid challenging the prior appropriation doctrine of water and its associated preemptive range rights.

WHAT KIND OF PROPERTY IS WATER?

If water is property, and we have seen where the courts have recognized it as property; just what kind of property is it? Wells Hutchins provides an answer in his *Water Right Laws in the Nineteen Western States* quoted in an earlier chapter:

—The appropriative right is a right of private property. It is subject to ownership, disposition, and litigation as in the case of other forms of private property.

—The appropriate right is valuable property.

—The general rule in the West is that the appropriative right is real property. . . .

—The appropriative right is an interest in real estate. As a general practical matter, a person who is legally competent to own title to land in a particular jurisdiction has equal competence to hold title to an appropriative water right therein.[20]

The appropriative water right is an interest in real estate. However, it differs from land title in that it is a right of use and is subject to loss by non-use. Even a temporary non-use forced by regulatory interference could raise the taking issue.

The *Attorney General's Guidelines for the Evaluation of Risk and Avoidance of Unanticipated Takings* cast light on this discussion:

Regulatory Takings

a. Like physical occupations or invasions, regulation which affects the value, use, or transfer of property may constitute a taking if it goes too far. *Pennsylvania Coal* v. *Mahon*, 260 U. S. 393 (1922); *Hodel* v. *Irving*, 107 S. Ct. 2076 (1987); *Nollan* v. *California Coastal Commission*, 107 S. Ct. 3141 (1987). Regulation has gone too far and may result in takings liability if:

> i. The regulation in question does not substantially advance a legitimate government purpose; it is not enough that the regulation or action might rationally advance the purpose purported to be served; or

> ii. In assessing the character of the government action, the economic impact of the action on the property interest involved, the extent to which the regulation conflicts with the reasonable, investment-backed expectations of the owner of the property interest, and other relevant factors, justice and fairness require that the public, and not the private property owner, pay for the public use. *Pennsylvania Coal* v. *Mahon*, 260 U. S. 393 (1922); *Penn Central Transportation Company* v. *New York City*, 438 U. S. 104 (1978); *Agins* v. *City of Tiburon*, 447 U. S. 255 (1980); *First English Evangelical Lutheran Church of Glendale* v. *Los Angeles County*, 107 S. Ct. 2378, 2389, n. 10 (1987).

b. Regulatory actions that closely resemble, or have the effect of, a physical invasion or occupation of property are more likely to be found to be takings. See *Nollan* v. *California Coastal Commission*, 107 S. Ct. 2076 (1987). The greater the deprivation of use the greater the likelihood that a taking will be found.[21]

Wilderness legislation, while specifically singling out grazing and water rights as protected interests within wilderness boundaries, sometimes imposes regulations so limiting that grazing becomes economically unfeasible.

THIRD PARTY INTERFERENCE
In regard to grazing permits issued by the BLM or Forest Service the general rule is that the agency has no more right to interfere with permit use than does any other third party.

> While grazing permits on government land are unrevoked, neither grantor nor third party may interfere with their exercise and government has affirmative obligation to safeguard them adequately.[22]

Again, from the attorney general's guidelines on permits granted by the government:

> The programs of many agencies require private parties to obtain permits before making specific uses of, or acting with respect to, private property, without necessarily effecting a taking for which compensation is due.[23]

A grazing permit, for example, is required for the use of private water and range rights on the federal lands. "Those agencies may place conditions on the granting of such permit."[24]

The typical grazing permit is conditional upon the specific period grazing can occur. This condition, purportedly, is applied to protect the forage from unseasonable use. This type of condition on private water and range rights is not necessarily a taking for which compensation is due.

However a condition on the granting of a permit risks a taking implication unless:

a. The condition serves the same purpose that would be served by a prohibition of the use or action; *and*

b. The condition imposed substantially advances that purpose.[25]

In the conditional season-of-use example, the condition imposed to prevent unseasonable forage use would have to serve the same purpose as the prohibition of grazing in that same time period. If it did not serve the same purpose as prohibition of grazing, a taking could be implied. This immediately gives rise to questions about BLM and Forest Service permit conditions designed not to prevent all grazing for the off-season but to provide forage for uncontrolled growth in wild horse and burro populations or the introduction of non-indigenous wildlife species to the rancher's range and water rights.

A season-of-use condition substantially protects the forage from possible overuse at the wrong time of the year and under those circumstances has no taking implication. If the effect of the conditional season-of-use only creates forage for competitive grazing by other species, according to the Guidelines this action risks a taking implication.

DELAY

A common occurrence when stockmen face regulatory action by the BLM and Forest Service is extensive delay on the part of the agency before decisions are rendered. The effect of this delay is often intended to prevent the stockman from making needed adjustments in his operational procedures to comply with the uncertain outcome of the agency proposed action. This can result in substantial added costs to the stockman.

The Guidelines say about delay when a federal agency implements a regulatory policy or action:

Undue delay in decision making processes, whether intentional or unintentional, may give rise to takings liability, or increase the amount of compensation due if the decision making process conflicts with the use of property pending the decision.[26]

Do the guidelines address the peculiar situation on the federal lands which we have identified as a split estate? In assessing the character of the government action, the Guidelines say, an agency should examine the extent to which the intended policy or action totally abrogates a property interest which has been historically viewed as an essential stick in the bundle-of-property-rights.[27]

For those persons or groups which do not recognize the rancher's interest in the range as *property* even though the rancher has substantial investment in the range, the Guidelines say:

> To the extent reasonably possible, an agency should examine the degree to which the proposed policy or action will interfere with reasonable, investment-backed expectations of those private property owners affected by the proposed action, even if such expectations are not formally recognized as property interests under the generally applicable law.[28]

The use of regulatory procedure to increase the cost of operation for the stockman on federal lands is also addressed in the Justice Department Guidelines. A "Takings Implication Assessment" (TIA) must include the identification and consideration of alternatives, if any, to the proposed policy or action which also achieves the government's obligation under law but would reduce intrusion on the use or value of private property.[29]

The logical conclusion one reaches after review of the taking issue is that if a rancher's rights and possessory interests in these federal lands are property, then much federal policy toward the stockman raises the taking issue.

In *McNeil* v. *Seaton*, on June 16, 1960, the Circuit Court of Appeals for the District of Columbia held that in the Taylor Grazing Act the stockman definitely acquired rights whether they be called rights, privileges or bare licenses, or by whatever name, while they exist they are something of real value which have their source in an enactment of Congress.[30]

THE THIRD PARTY

We have spent much of this book establishing the difference between the grazing right, which had its origin in local law and custom and court decisions, and the grazing permit which was originally granted by the government in recognition of the pre-existing right of the rancher on the federal grazing lands. The permit, a privilege granted by the government has been held by the courts to be revocable. The underlying property rights established under the prior appropriation doctrine of the states was not created by the federal government and is not revocable.

A direct confrontation over the underlying grazing right issue, as a vested interest in the federal lands themselves, has been consistently sidestepped by the federal government for almost one hundred years, especially since the ruling in 1907 in the case of *Kansas* v. *Colorado*.[31] The court in that case smashed the Forest Service contention that they could defy the prior appropriation doctrine of the states. Since then, federal land policy has been influenced by a fear of what the courts would say if a well structured case on grazing rights were to reach the Supreme Court.

The best challenge was seen, perhaps, in *Hunter* v. *United States*. In that case the court partly avoided the grazing rights issue by citing the fact that

Hunter's water could feasibly be transported to another site for grazing purposes.

The implication of an affirmative answer to the grazing rights issue are enormous. Vast wealth was stripped from the western states and their citizens by the Forest Reserve Act of 1891. If the courts were to affirm, at this late date, the validity of the preemptive rights of ranchers, then damage claims by western citizens for the extortion worked upon them for almost one hundred years would be substantial, to say the least. The circumstances are similar to the property claims of certain Indian tribes which the government had violated by ignoring legitimate rights held by the Indians. Some Indian tribes have won substantial damages from the federal government. An affirmation by the courts that preemptive rights to the range do in fact exist, and have existed since before March 3, 1891, could bring unprecedented demands for restitution of lost values. A political decision by the courts, not a strict legal opinion, could well be the method by which such a confrontation may be resolved.

The incessant attack by the federal agencies and the environmental movement on the stockmen in recent years makes a confrontation over the property rights issue virtually unavoidable. The stockman is rapidly being pushed into the corner where he has everything to gain and nothing to lose by such a confrontation.

Partially as a result of the desire to avoid a direct answer to the pre-existing grazing rights issue, and partially as a result of BLM and Forest Service efforts to settle true grazing right issues before they can reach the federal courts, case law on the topic is not entirely clear.

As regards the permit to graze, based on these original rights, case law appears to give a more precise direction. Case law has repeatedly asserted that the permit to graze granted by the government is revocable and non-compensable. However, while it remains unrevoked, it is something of value and the federal agency has an obligation, not only to not interfere in its use, but is obligated to defend that use from interference by third parties.[32]

In *McNeil* v. *Seaton* the court held that the Secretary of the Interior had an affirmative obligation to protect grazing rights.

In *Hatahley* v. *United States* agents of the federal government attempted to impound animals belonging to Bill Hatahley without proper conformance to the rules relating to trespass. The federal agency settled for monetary damages paid to Hatahley.

In *United States* v. *Fuller* the U. S. Supreme Court held that while grazing permits on government's land are unrevoked, neither the grantor or a third party may interfere with their exercise and that the government has an affirmative obligation to safeguard them adequately.

In *Red Canyon Sheep Company* v. *Ickes* the court held that the purpose of the Taylor Grazing Act was to provide for the most beneficial use possible of the public range in the interest of grazers and the public at large, and to define grazing rights and to protect those rights by regulation against interference. The court held further that a lawful business was property and that grazing on the federal lands is a lawful business.

In the 1944 case *Osborn* v. *United States* the Ninth Circuit Court of Appeals answered the Forest Service's long-standing contention that a grazing permit has no property value, a premise which the Forest Service had attempted to establish through regulation. The court held that a regulation that a grazing preference is not a property right does not mean that a permit is not a property right and cannot be so construed.

In the case of *Oman* v. *United States* rancher Odell M. Oman had purchased certain grazing rights and was granted a permit for their use by the federal agency (BLM). Personnel of the agency then allowed another livestock operator to run livestock on Oman's permitted grazing lands in conflict with the rights acquired by Oman. Oman contested the agency action in the courts and the court said that where there was no attempt to revoke but instead an outright interference with plaintiff's grazing rights by servants of the government while their permits remained outstanding and unrevoked, the United States through its agents had no more right to interfere with their exercise than would any third party. Grazing permits granted under the Taylor Grazing Act confer rights subject to judicial protection including an action at law for damages for an interference therewith.[33]

The courts have been more precise in defining the relationship of the federal government as grantor of grazing permits and the stockman as grantee as they relate to any interference by a third party. In *Oman* v. *United States* the interfering third party was clearly identified as an agent of the federal government.

Again we must ask the question: In light of attempts to replace livestock grazing taking place under unrevoked grazing permits, with competitive uses such as recreation or wildlife intrusion, has there been third party interference by a government agent? When ranchers are pressured to the point of extortion to accept reduced grazing capacity for the accommodation of federally managed wild horses and burros has there been third party interference by agents of the government?

When the agency endorses a range management policy based on political considerations, to the extent of violation of moral and scientific standards to accommodate competitive interests in conflict with the use and enjoyment of an unrevoked grazing permit, has there been intentional third party interference? I refer to instructions to Forest Service range administrative trainees given at a training session in 1956: "We should be morally right and we should be scientifically correct, but neither is the controlling factor in getting the job done. The controlling factor is public opinion."[34]

The Internal Revenue Service has a different policy. In a letter to the Forest Service dated August 25, 1988, an IRS attorney for estate taxation wrote:

> We are aware of the Forest Service policy that National Forest Service grazing permits have no value. However, the Federal tax statutes require us to value the assets of a decedent or donor based on their "fair market value." Even though you may not place value on the leases, the ranchers do whenever they buy or sell a ranch. Every ranch with substantial leases sells on an animal unit basis considering the grazing leases, and every reputable fee appraiser places a value on them despite your regulations.

For the Internal Revenue Service to value the leases does place government agencies in an inconsistent position. But the greatest inconsistency is for the taxpayers to rely upon your regulations to claim the grazing rights have no value for tax purposes, and then to universally place a value on them when they sell or mortgage their ranches.

It would appear that the market place is the best indication of whether or not there is a market value for the leases. As long as the market place says there is value, a fair evaluation of a ranch cannot be made without considering the price a willing buyer would pay for the grazing leases.[35]

The discussion in this chapter leads to three major questions about federal grazing land policy: (1) Have the BLM and Forest Service failed to adequately protect the viability of unrevoked grazing permits? (2) Have the BLM and Forest Service intentionally interfered on behalf of third parties in the use and enjoyment of the stockmen's grazing rights? (3) Has it been a de facto policy, engaged in by the BLM and Forest Service, to *take* the rancher's property right interests in the federal lands?

Chapter Twelve Footnotes

1. See for example *Mollohan* v. *Gray*, 413 F. 2d 349. (Decided June 17, 1969 by the Ninth Circuit Court of Appeals); *Acton* v. *United States*, 401 F. 2d 896. (Decided October 4, 1968 by the Ninth Circuit Court of Appeals); *United States* v. *Cox*, 190 F. 2d 293. (Decided June 21, 1951 by the Tenth Circuit Court of Appeals).
2. *First English Evangelical Lutheran Church of Glendale* v. *County of Los Angeles.* 107 S. Ct. 2378. (1987).
3. *Nollan* v. *California Coastal Commission*, 107 S. Ct. 3141. (1987).
4. Executive Order No. 12630, "Governmental Actions and Interference With Constitutionally Protected Property Rights," *Federal Register*, vol. 53, No. 53, p. 8859, Friday, March 18, 1988.
5. Ibid., p. 9.
6. *Armstrong* v. *United States.* 364 U.S. 40, at 49. (1960).
7. *Attorney General's Guidelines for the Evaluation of Risk and Avoidance of Unanticipated Takings*, unpublished internal document, U. S. Department of Justice, Washington, D. C., 1988, p. 11.
8. *Hunter* v. *United States.* 388 F. 2d 148, at 149. (1967).
9. Ibid.

10. Ibid.
11. *Attorney General's Guidelines for the Evaluation of Risk and Avoidance of Unanticipated Takings*, pp. 11-12.
12. Ibid. p. 12.
13. Hutchins here cites Nev. Rev. Stat. ## 533.485 to .510 (Supp. 1967).
14. Hutchins here cites N. Mex. Stat. Ann. ## 75-1-4, 75-1-5, and 75-8-3 (1968).
15. Hutchins here cites Oreg. Rev. Stat. # 537.300(2) (Supp. 1969).
16. Hutchins here cites S. Dak. Comp. Laws Ann. ## 46-1-6(3) and 46-4-1 to 46-4-8 (1967).
17. Hutchins here cites Tex. Rev. Civ. Stat. Ann. art. 7500a (Supp. 1970).
18. Hutchins here cites *Steptoe Live Stock Co.* v. *Gulley*, 53 Nev. 163, 171-173, 295 Pac. 772 (1931).
19. Hutchins, *Water Rights Laws in the Nineteen Western States*, vol. 1, pp. 538-539. I have silently removed irrelevant references.
20. Ibid. p. 443.
21. *Attorney General's Guidelines for the Evaluation of Risk and Avoidance of Unanticipated Takings*, pp. 13-14.
22. *United States* v. *Fuller*. 442 F. 2d 504. (1971).
23. *Attorney General's Guidelines for the Evaluation of Risk and Avoidance of Unanticipated Takings*, p. 15.
24. Ibid.
25. Ibid.
26. Ibid., p. 16.
27. Ibid., p. 18.
28. Ibid., p. 19.
29. Ibid.
30. *McNeil* v. *Seaton*, 281 F. 2d 931. (1960).
31. *Kansas* v. *Colorado*, 206 U. S. 46. (1906).
32. *Oman* v. *United States*, 179 F. 2d 738. (1949). *McNeil* v. *Seaton*, 281 F. 2d 931. (1960). *Hatahley* v. *United States*, 351 U. S. 173 (1956). *United States* v. *Fuller*, 409 U.S. 488. (1973). *Red Canyon Sheep Co.* v. *Ickes*, 98 F. 2d 308. (1938). *Osborn* v. *United States*, 145 F. 2d 892 (1944).
33. *Oman* v. *United States*, 179 F. 2d 738 (1949).
34. 1956 Ranger's Training Camp, "Range Administrators," May 7, 1956, Box 3, 13/2/1:3, Record Group 95, Federal Records Center for the National Archives, Denver, Colo.
35. Richard P. Chamberlain, Internal Revenue Service, to R. B. Tippeconnic, Forest Supervisor, Coronado National Forest, Tucson, Arizona, August 25, 1988, IRS correspondence code number 1235-4208PX. In the files of Coronado National Forest. Letter bears a Forest Service "Received" stamp dated Aug. 26, 1988.

EPILOGUE

In his book *Sectionalism and American Political Development:1880-1980,* historian Richard Bensel addressed current northern core influence on the economy of the western states in light of environmental policies of the federal government:

> [M]ost major environmental policies directly retard the development of natural resources in the West and, thus, indirectly stimulate the exploitation of similar resources (where they exist) in the East. Strip-mining restrictions, leasing for off-shore drilling on the continental shelf, *and legislated prohibitions on the private use of western land in the public domain all fall into this category.* The classic example of this kind of policy has been the scrubber requirement imposed by the Environmental Protection Agency on low-sulphur, western coal (for which scrubbing is unnecessary under most standards) in order to stimulate the processing of high-sulphur eastern coal. Since the cost of installing scrubbers for "dirty" eastern coal would otherwise force utilities to abandon its use, the EPA requirement has rescued an otherwise moribund energy source, albeit at a substantial cost in economic efficiency.[1] [Italics added.]

The attitude of the northern core toward the West had been expressed in terms of colonialization of the western states by taking their land base from them in various reserves. The West's resources were first plundered from its pioneers by eastern capitalists attempting to usurp the prior appropriation benefits of the Act of 1866, then under the total lockout of the Forest Reserve Act of 1891, later by the Taylor Grazing Act, and still later by the Federal Land Policy and Management Act.

The forest reserves were imposed on the West amid a din of conservation rhetoric which effectively masked the true intent of northern core interests. As long as the conservation view was needed, conservationists obtained considerable funding from the northern core capitalists.

The Taylor Grazing Act again prevented full recognition of preemptive range rights by the expansion of another bureaucracy to manage the federal split estate. This too was brought about amid the conservation fervor that accompanied the Roosevelt administration's dramatic turn to the left.

The Federal Land Policy and Management Act was essentially *designed and manufactured* by the conservation movement now wearing the title "environmentalism."

The turn of the century conservationists had the roots of their support in the northeastern quadrant of the country. The result of their effort was to restrict the political economy of the West in favor of northeastern financial and political interests. The conservationists of the 1930s had the roots of their support among the northern core interests who put New Yorker Franklin Roosevelt in office.

223

The broad effect of the Taylor Grazing Act was to greatly increase the amount of territory held in a quasi-colonial status for the benefit of the same northeastern financial, industrial, and political interests.

Among the major contributors to the environmental movement of the 1970s and '80s are found the interests of the Rockefellers, Fords, Mellons, Atlantic Richfield Corporation, and many other well known entities which are part of the present financial and industrial fabric of the northeastern United States and, in many cases, the heirs of their 1890s counterparts.

PHILOSOPHY AFFECTS POLICY

At the turn of the century the conservation movement emerged in two separate branches. The nature preservation branch, represented most visibly by John Muir and the Sierra Club, and the utilitarian "wise use" branch represented by Gifford Pinchot and his followers.

This bifurcation of the conservation movement is still present in the 1980s. It is difficult to write about the underlying philosophical or religious beliefs of any politically powerful faction because such critical analysis can be easily mistaken for religious prejudice or mere name-calling. However, as Rutgers University historian Susan R. Schrepfer once noted, historical narration often neglects the fact that the politics of individuals and groups reflect their philosophic assumptions and that, when beliefs change, behavior changes.[2] My comments here are intended only to show that specific beliefs have specific policy consequences.

In a comparative study of two nature preservation groups—the Sierra Club, founded in 1892 and the Save-the-Redwoods League, founded in 1918—Schrepfer noted that both shared a single philosophical and religious outlook that shaped their political actions. It is worth quoting the historian at length:

> The Sierra Club had been formed in 1892, during a period when many sought to reconcile theism and Darwinian evolution. One Club founder, Joseph LeConte, a University of California paleontologist, outlined for Americans a synthesis of direct agency and natural law. LeConte's world view was predicated upon the existence of a force external to the material world and upon man's ability to grasp the meaning of this force. He advanced a rational, scientific theology in which God has both a real, independent existence and a presence in nature. LeConte's idealism saw man as possessing two natures—the animal and the spiritual. With these dualities, LeConte perceived design in nature. Evolution was creation, "the conception of the one infinite, all-embracing design, stretching across infinite space." The very existence of man proved that evolution moves toward greater physical and spiritual development.[3] From the late nineteenth century through the 1930s, such Sierra Club officers as naturalist John Muir, business leader Duncan McDuffie, professor of theology William Badè, and Stanford University President David Starr Jordan similarly influenced the organization with their conceptions of progressive designs in nature.[4]

When the Save-the-Redwoods League was established in 1918, its leaders also held this vision of a moral universe, despite the ready evidence of global evil. John Campbell Merriam, paleontologist and League president, preached reconciliation of the religious impulse and evolution. To Merriam research was an act of reverence. Naturalists and League officer Henry Fairfield Osborn and Vernon Kellogg similarly wrote that science was verifying the existence of a central, moral wisdom in the universe.[5] Scientists were proving not only that the world was progressive but that it was one in which man could control his destiny. If the human intellect is the highest product of directional evolution, our mastery of the world must be beneficial. Man's control of the biological world was progressing with amazing rapidity.[6]

This philosophy of a universe with teleological purpose and direction had important policy consequences: If man was able to grasp the design in nature, then parks should be temples in which to worship a universe of order and design. Therefore, these men designed their parks to capture *only* the highest examples of evolutionary momentum, such as the glacier-carved valley of Yosemite and the oldest coastal redwoods. Every worshipper was welcome: One did not lock the cathedral door during services.

Such men also respected technology. As Schrepfer commented, "Man had the right and ability to manipulate nature for his benefit. Parks were to be laboratories that would aid this process." The eastern capitalists and their money were welcomed into the organizations and many joined and contributed huge sums, such as William Harkness and John D. Rockefeller, Jr.

However, things changed. In the 1950s and 1960s a philosophical naturalism had replaced the teleological view. No longer was evolution an expression of God or a creator. The universe no longer had purpose or direction.

The policy consequences of this change in philosophical beliefs were remarkable. Sierra Club naturalists such as Loren Eiseley initiated a campaign for an Earth National Park. The universe and nature were no longer man's benevolent home welcoming his manipulations. The earth was the source of all man's needs, even those he might be unaware of, and man could not escape either his needs or the earth. Now the earth was man's prison, physical and spiritual.[7] The teleological view that urged man to move beyond the physical order to grasp the spiritual order had been abandoned. As Schrepfer wrote, "The vision had been lost."

[E]mpirical evidence supported only the belief that randomness characterized the universe. The natural sciences no longer offered proof that man's brain and technology were the summit of the momentum of evolution. Biology revealed interdependence, specialization, and adaptation, but allowed no judgment as to superiority of species. Man was an accident of the chance-filled course of genetics.[8]

Inherent in this philosophical naturalism was also the conviction that since man can never escape the cosmos neither has he mastered it. To speak of man, trapped in his body and limits of time and space, conquering the mysteries of the universe was, according to Eiseley, "the equivalent of installing a grasshopper as Secretary General of the United Nations."[9] To the Club, wilderness held answers to more questions than man yet knew or might ever know how to ask. The role of the scientist was to understand the world, not to seek to master it.[10] Eiseleys's random universe denied man a world priority. Man was a temporary adaptation to a specific environment and not innately superior to other species. To many Club leaders of the 1960s, nature had rights that should not be continually compromised for the benefit of man.[11]

Historian Schrepfer performed this analysis to explain the split between the Sierra Club and the Save-the-Redwoods League during the fight over Redwood National Park in the 1960s—the League had remained true to the vision but the Sierra Club had abandoned it. The two old friends became political opponents and recommended different places and different measures for Redwood National Park.

We may gather our own meaning from Schrepfer's work. The new leaders of the 1950s and 1960s Sierra Club lived in a universe of no purpose and no direction. Their fellow man was not the highest expression of evolutionary momentum—even the notion of God as evolutionary creator had been lost. Club leaders rejected technology. Belief in progress was rejected. Man no longer had the right to manipulate nature for his benefit. Parks were definitely not to be laboratories that would aid progress. If man had an expansionary urge to manipulate the earth or "run laughing among the stars" as William Blake poeticized, then it must be only nature's programming, a means of spreading, as with spores, a disease. Man was now the destroyer of all things bright and beautiful, and no longer could claim any part of that brightness and beauty for himself. Now every square inch of the earth was to be saved—from man.

This new policy had new political consequences. The eastern capitalists were turned out of the Sierra Club, but their money remained welcome: the Rockefeller brothers funds, for example, gave many millions during the 1960s and 1970s. No longer would the Club lobby for the best examples of evolutionary progress, but would lobby against man's use of the earth *in toto*. Now there were no limits to what should be saved, which gave limitless scope for the Club's fundraisers. Earth should be a gigantic national park. Nothing man-made was immune from the Club's political attacks.

The nature preservationists, including all of the top ten organizations of today, have evolved into "whole earth" environmentalists. All prior sense of discrimination in preservation vanished. New organizations joined the old line groups such as the Audubon Society (1905), the Izaak Walton League (1922), the Wilderness Society (1935), and the National Wildlife Federation (1936). Now came militant and strident activists in the Environmental Defense Fund (1967), the Friends of the Earth (1969), the Natural Resource Defense Council (1970), and all the rest. Lawyers and other professional leaders replaced the old amateur naturalist cadre as heads of environmental organizations. Skilled professionals at the movement's helm hired public relations and commercial fund raising firms to maximize their political clout and financial power. The former citizen movement gathered to itself all the accoutrements of the northeastern core corporate establishment: Offices in Washington, D. C. close to the Capitol, public opinion researchers, telemarketing bull pens to recruit members, staff lobbyists, media departments, a combined annual budget for the top ten groups in excess of $200 million. The Sierra Club even advertises its mailing list for rent to commercial firms, touting these demographics: 57.3% Male, 42.7% Female, 51.1% Married, 40.3 Median age, 96% Attended college, 34.3% Earn $50,000+, $40,300 Median household income, 87.8% Hold professional managerial jobs. The era of corporate environmentalism emerged with the 1970s.

The northeastern core capitalists continue to heavily fund the environmental movement. Among the better known capitalist supporters of corporate environmentalism, according to Sierra Club publications, are Atlantic Richfield, BP America, Chemical Bank, Corning Glass Works, CPC International, Digital Equipment, Equitable Life Assurance, Federated Department Stores, Georgia Pacific, IBM, John Hancock Mutual Life Insurance, Levi Strauss, R. H. Macy & Company, McGraw-Hill, Pepsico, Pfizer, Phillip Morris, Pitney Bowes, Sara Lee, Squibb, TransAmerica, Union Pacific, United Technologies, Wells Fargo Bank, and Westinghouse Electric.[12]

The movement also became expert at securing federal funding for their private environmental corporations. From 1974 to 1982 the National Wildlife Federation obtained a total of $915,377 in federal grants. The Sierra Club got $790,746 from 1978 to 1981, the National Audubon Society got $221,483 from 1977 to 1981, the Environmental Defense Fund $96,131 from 1978 to 1981, the National Center for Appropriate Technology $3,828,885 from 1978 to 1981, the Center for Renewable Resources $1,309,540 from 1979 to 1981, and the Natural Resources Defense Council more than a million dollars from 1978 to 1982.[13]

The radical eco-terrorist branch of the environmental movement operates a sabotage and arson program intended to destroy livestock facilities and bankrupt ranchers. Earth First! took credit for burning a privately-owned California livestock auction building in 1988. Cattlemens' association buildings in the Southwest have been defaced and the lives of association executive directors threatened by Earth First! Numerous fence removals, stockwater facility vandalisms and livestock poisonings have been claimed by Earth First!

Free Our Public Lands! is an Earth First! project based in Cottonwood, Arizona. Lynn Jacobs, an Earth Firster listed as a contact for Free Our Public Lands!, wrote a 48-page tabloid describing private use of federal grazing lands as "welfare ranching," an emotional slogan designed to divert attention away from the split estate ownership of federal grazing land. Free Our Public Lands! sells bumper stickers reading "Get Livestock Off Our Public Lands!" and "Protect Our Environment—Stop Public Lands Livestock Grazing!"[14]

The utilitarian "wise use" branch of the conservation movement also survives in the 1980s. The coinage of Gifford Pinchot in 1907 that "conservation is the wise use of resources" still informs a large professional and citizen organization movement.[15]

However, "resource protection" is also the watchword of official government environmental policy makers who have their own power agenda. But as historian Richard Bensel points out, government conservationists use resource protection as the stalking horse behind which sectional political goals are pursued.[16]

Environmental sanctions are generally applied more intensely in the West and Southwest than in other portions of the country. Political environmentalism, to achieve ends beneficial to the northern core, is behind the ongoing environmental harassment and economic vandalism the rancher faces on the federal range.

Until property rights are clearly defined, wasteful competition for the control of the western federal lands will continue. Throughout the controversy the preemptive right of the rancher remains unperceived, as if it were a total enigma. While the northern core interests have prevented formal blanket recognition of that right, for almost one hundred years that right has never been fully erased. One hundred years of repeated assaults on the preemptive range right have served more to strengthen the validity of the right than to weaken it. The rancher's claim to a pre-existing right to graze, based on the prior appropriation doctrine of the western states, still promises to rear its head in a decisive manner despite the political, legal, and public relation attacks which have been mounted against it since the 1890s.

The lands themselves are becoming more valuable. The population center of the nation is moving toward the West and Southwest. The more the population shifts to the West and Southwest, the greater the clamor for a solution to the federal lands problem.

With population growth comes political power. This too has been shifting westward. While northern core political interests are still able to heavily influence the western states' election process, western states are nevertheless increasing in political clout.

While Sierra Club and sister environmental organizations may lobby and campaign successfully for northern core interests in the western states, growing economic and political reality favors a West in control of its own destiny somewhere in the near future. This will demand a solution to the federal lands problem.

SOLUTIONS

The solution that environmental organizations have pursued is to clear title on the western lands in favor of the federal government by creating an economic-political climate that precludes the survival of the rancher, miner, logger, and other claimants of the split estate.

This approach, if pursued, would most likely raise a massive legal challenge—a prospect that cannot be ignored. The *de facto* policy of federal land management agencies to take lawful property without compensation behind a facade of "protecting" resources invites a major Fifth Amendment challenge which could raise damage claims of monstrous proportions.

Other solutions have been offered by recreational and environmental interests, as well as stockmen, proposing a market-oriented approach to the range rights question. Appraise the range at its current fair-market value, they suggest, and buy the rancher's interest through a cooperative transaction, or if necessary, condemnation.[17]

This solution seems not only logical, but fair. However, federal policy has blocked all such efforts. Recognition, it is argued, would be an acknowledgement of the pre-existing right which was the basis for the desperate action expressed in Section 24 of the 1891 Act. It would be difficult to confine payment for the rancher's range rights to the current fair-market value of grazing. Over the past one hundred years literally trillions of dollars worth of resource wealth has been stripped from the western lands to support the northeastern infrastructure. Any affirmation today that the preemptive range right is valid would be an admission that it had been valid since before 1891. Damage claims against the federal government for the plunder of this resource wealth could be enormous.

An acknowledgement that the rancher's pre-existing right has been valid since 1891 would also raise the issue of title as it affects current non-grazing activities on the federal lands. Conflicts over who rightly owns the oil wells, mines, timber stands, hydro power projects, recreation sites, and other commercial activities on these lands would be raised.

In other words, we have come full circle in the last one hundred years. A desperate effort in 1891 to subjugate the West with forest reserves, followed by other restrictive legislation in the 1930s and 1970s, while serving to transfer massive wealth to the northeastern quadrant of the nation, has not really answered the question of title.

Let's examine further what some of those questions of title are. If the courts were to clearly decree the preemptive grazing right invalid, this would, of necessity, invalidate the tens of thousands of stock water rights based on prior appropriation with livestock grazing as proof of beneficial use. Such a decree would have to be a unilateral action by the courts in total defiance of the entire prior appropriation doctrine of water law which rules the western states. If livestock grazing—which has been recognized by *local law and custom and court decisions* as a lawful proof of beneficial use to establish legal title to water on the federal split estate—can be arbitrarily decreed to be an unlawful action by the federal government, then is any of the prior appropriation doctrine safe?

If stock water implies a use of the adjacent grazing area as an essential factor in the use and enjoyment of the right, (and certainly the water base permit demonstrates this essentiality) how can it be said there is no right in the land itself? If the federal government were, through the courts or Congress, to clearly abrogate the range right, it could as easily abrogate the irrigation water right, the mining water right, municipal water rights, and all the private property that was created from the federal lands as a result of putting those waters to use under the prior appropriation doctrine over the past century and a half. In other words an attempt to clearly nullify the preemptive range right could open the door for a challenge to virtually all the property rights, rural or urban throughout the West.

The stockman operating on the western rangelands today finds himself thrust unsolicited into the position of "the spoiler." The federal split estate lands have increased dramatically in value over the past one hundred years. Population pressures continue to grow. Political strength of the western states is increasing in proportion to population growth. Use demands on the federal split estate increase in proportion to the population growth.

A recent major effort to disenfranchise the western stockman from the range came through an action on the part of both the BLM and Forest Service to assert a federal non-reserved water right that would be of a date prior to the rights of the stockmen. The federal position was expressed by Interior Solicitor Leo M. Krulitz in June of 1979.[18] In a solicitor's opinion known simply as "the Krulitz opinion," Krulitz interpreted the Executive Order of April 17, 1926—known as Public Water Reserve Order No. 107—to support federal claims to a host of waters that were privately owned.[19] A major conflict with permit holders on BLM and Forest Service grazing allotments ensued.

The Krulitz opinion was repudiated in September, 1981, by Krulitz's successor William Coldiron. Coldiron accepted the extensive body of law and court opinion on the subject of western water rights. "He concluded there was no such thing as a federal non-reserved water right. He further went on to say that Federal agencies must acquire water as would any other claimant within the various states even though the water was on federal lands."[20]

The general policy of Interior Secretary James Watt to honor states' water rights and the prior appropriation doctrine caused profound alarm in the environmentalist community and its northeastern core supporters. Bureau of Land Management Director Robert Burford developed a stockwatering policy in line with states' rights. Environmentalist and northeastern core concerns over the Watt-Burford policy were expressed in a secret 1984 congressional report suppressed by the House Committee on Appropriations:

> This policy has serious implications for multiple-use management of the public lands because it allows single-use interests, e.g., livestock grazing, to control use of water on these lands. (Page 8)
>
> As one BLM official put it, "the beneficial use is to the permittee—BLM in reality is co-holder of nothing." (Page 10)
>
> Owning the water would give a rancher a "vested interest" in the land—something western grazing permittees have been seeking for generations. (Page 11)
>
> The permittee becomes insulated from BLM management or land use changes while being able to make those changes himself. (Page 11)
>
> In some States a water right becomes a *private property right* once in the name of the permittee. A vested interest in public lands and its resources goes into private hands. (Page 11)
>
> [The BLM State Director in California said] "What recourse will BLM

have as landowner if the permittee sells or transfers his water right? According to past water law decisions, once an individual has water rights, he can do as he pleases with the water so long as the use is beneficial according to the State law." (Page 12)

The underlying problem would become one of a vested interest in the public land (water being a scarce and essential resource) being held by a private user with a single-use interest. (Page 13)[21]

It is hardly surprising that the House Appropriations Committee Chairman has chosen to permanently suppress this report. It states bluntly the power of private rights, if recognized, to loosen the grip of the Forest Service and the Bureau of Land Management over lawful owners of prior appropriation water. The Watt-Burford policy forced federal agencies to recognize the plain wording of Section III of the Taylor Grazing Act:

That nothing in this sub-chapter shall be construed in any way to diminish or impair any right to the possession and use of water for mining, agriculture, manufacturing or other purposes which has heretofore vested or accrued under existing law validly affecting the public lands on which may be hereafter initiated or acquired and maintained in accordance with such law.[22]

Stock water rights had their origin in the prior appropriation doctrine of the states. They were pre-existing rights. Most water rights pre-dated the Taylor Grazing Act and the 1891 Forest Reserve Act. Most of these stock water rights pre-dated the repeal of the preemption laws. These rights represented the "right, title, interest, or estate" that the rancher owned by virtue of the prior appropriation doctrine.

These rights were not acquired through the "creation of a grazing district or the issuance of a permit." These rights had existed prior to the creation of grazing districts. These rights had existed prior to the issuance of permits. There is a clear distinction between the underlying pre-existing right created through *local law and custom and court decisions* and the revokable, non-compensable permit granted by the federal government.

RANCHERS IMPROVE THE RANGE
Historically, range rights have served the environment in ways that generally go unappreciated. The myriad water developments installed by stockmen in the arid regions have stimulated a vast expansion of wildlife habitat and recreational areas. The availability of stock water allowed wildlife to thrive and increase. Predator control measures that stockmen employed to protect domestic livestock also stimulated the expansion of wildlife populations.[23]

The history of the Great Basin is instructive of the relationship between livestock grazing and wildlife populations. Early explorers in the Great Basin recorded the almost total absence of game animals. Sightings of mule deer were few. The indigenous Indians in the Great Basin were derisively referred to as "digger" Indians, a reference to their bare subsistence lifestyle heavily dependant upon digging for edible roots. A primary source of protein was grasshoppers and ants. It can be safely assumed these Indians did not subsist on grasshoppers and ants because they did not like venison. They did not eat venison because deer were not present in most of the Great Basin.[24]

One hundred years after the first explorers traversed the Humboldt, some forced to eat their horses and mules and occasionally a coyote to survive, the Great Basin was host to a massive population of mule deer and renowned in sportsmens' circles as a premier hunting area. What had happened during that one hundred years?

A major reason desert areas tend to be devoid of most ungulate game animals such as deer is a simple matter of water. Where water sources are few and widely scattered deer cannot thrive in large numbers. A predator, on the other hand, can patrol those few water places with the assurance that all living things must come within his reach eventually. The vast expansion of watering places as stockmen vied for control of the range resulted in the addition of thousands of new water sources for the benefit of wildlife as well as livestock.

An 1866 map of the Great Basin prepared by the Interior Department identifies South Central Nevada as a "wasteland," "no wood, no water, no grass."[25] Today that area is well watered and one of the finest winter livestock ranges in the Southwest. The difference was made by over three hundred wells developed and maintained by stockmen.

Of even greater benefit to game herds in the Great Basin was the advent of the sheep industry. The introduction of sheep provided an easily available source of food for predators, reducing pressure on mule deer and other game species which tended, on occasion, to migrate from surrounding water-rich mountainous areas. By 1900 wool, lamb, and mutton production was a major industry in the Great Basin. Predator numbers had grown proportionately with the numbers of sheep, now their primary source of food. Predators consequently became a costly problem. By the time the United States entered World War I predator control was intensive throughout the Great Basin. Both government and private sources paid bounties and engaged in activities designed to hold predator numbers down.

In 1910 deer sightings were a rare phenomenon. By 1920 deer sightings in the Great Basin were commonplace. The 1930s saw deer hunting develop as a major seasonal activity throughout the region.[26] The 1940s and early 1950s brought Nevada into the forefront of mule deer hunting states. In 1955 mule deer numbers stood at an estimated 300,000, a peak level.[27] Domestic livestock numbers on the federal lands were estimated at 591,000 cattle and calves and 468,000 sheep and lambs, also a high level.[28]

From 1955 to 1965 there was a dramatic reduction in sheep numbers on the ranges of the Great Basin, caused by a combination of factors. Synthetic fabrics developed during the World War II period severely cut into the wool market. Post-war prosperity thus created a major problem for sheepmen. Federal officials continued to oppose the ranchers' preemptive right claims, targeting the sheepmen, who possessed most of the water-based grazing permits, for harassment.

By 1965 sheep had literally disappeared from many parts of the Great Basin. As sheep disappeared so did predator control. Predator numbers had adjusted to the increased availability of food from the sheep herds. Predators had nowhere to turn except to wild game. They turned on the mule deer and sage grouse populations with a vengeance. Subsequently, predator control programs were replaced with predator protection programs as the environmental movement began to demonstrate political clout and as a result, prey species declined further. By 1970 there were major concerns about the long term survival of mule deer and sage grouse in the Great Basin. The presence of stockmen had improved wildlife populations. Their disappearance harmed wildlife.

If it were not for the existence of the stockman's range rights, the federal lands would have been dramatically degraded by eastern capitalists' intensive development decades ago. If the waters on the western lands could have been seized by the federal government, as was the veiled intent of the Forest Reserve Act, it is doubtful if there would be any area of the West which would be undeveloped enough to meet the present day standards for wilderness. If the split estate in the federal lands could have been erased by the 1891 Act, subsequent legislation, and court rulings, then the West would not have remained an open, virtually unsettled area.

As we have seen, large corporations attempted to monopolize the West by numerous stratagems during the last quarter of the nineteenth century.* Their ultimate interest in this vastly rich natural resource storehouse was not to make a 600-million-acre playground for citizens of the western states. The original intent of these corporations and the current desire of their modern day counterparts was and is to control and profit by the eventual development of these lands. The controversy over the federal lands has never been "development versus non-development." The real issue has always been "who is going to develop it?"

While the stockman has been denied full title to the rangelands and thereby denied the right to enjoy their full economic potential, he has effectively prevented any other interest from exploiting the land to its full economic potential.

The rancher through his range right has created and maintained the recreational qualities of those vast open areas presently classified as federal range and forest lands.

DE FACTO RIGHTS

As noted in Chapter One, in the 1980s, even when the Reagan administration promised privatization and deregulation, performance lagged far behind policy. Ironically, it was the ranchers' split estate interest in the federal lands that scuttled the 1982 Reagan proposal to privatize by sale some 35 million acres, or 5 percent of the public lands. The Reagan administration discovered that you can't sell what you don't own. As the President's Commission on Privatization lamely admitted:

*Supra, p. 55-66.

A year later, in July 1983, the entire effort was unceremoniously canceled. It had foundered on the existence of de facto property rights. Over 50 years or more of continued use, ranchers had acquired a de facto property right to graze on particular parcels of public lands—rights even bought and sold in the market on occasion. As a practical matter and whatever the legalities, such informal rights cannot be taken from their holders without the government providing fair compensation. In such instances, the best strategy for achieving privatization may be to give (or to sell cheaply) to the holders of political property rights some portion—or conceivably even all— of the rights to the government property being divested. This strategy had not been followed in the 1982 planned sales.[29]

SUGGESTIONS

Is there a plausible solution to the dilemma of the split estate? Is there a way to unleash the full potential of our western lands without triggering gargantuan damage claims against the federal government for just compensation on the one hand, or the uncompensated destruction of the West's water laws and private property rights on the other hand?

Between these two unacceptable extremes I would like to explore the middle ground and propose the framework for a solution. My framework involves not the immediate elimination of the split estate, but a modification of it. At the same time it would set in motion the market factors which would sort out the competing and sometimes conflicting demands on the lands in accordance with public expectations and desire.

The primary cause of the problem and the primary barrier to a solution over the past one hundred years has been the preemptive range right of the rancher. No matter how vehemently agency heads and environmentalists deny its existence, it has still blocked their efforts to completely dominate these lands for over a century. Whatever the stockman's rights are, they need to be precisely defined and fairly dealt with.

Appraisal

The first step in this direction is to appraise the value of the rancher's lawful investment in the federal range under his control. This could be accomplished using standard appraisal methods following the same procedure used in determining the value of the rancher's interest in the split estate for estate tax settlements with the Internal Revenue Service.

There is already in place an official appraisal of rangeland values which was compiled in a joint BLM-Forest Service effort. The amounts listed in this appraisal are gross values. The rancher's share of the split estate could be subtracted from the gross figures in the appraisal to arrive at net residual values owned by the federal government. For example, if the appraised gross value were six dollars per AUM and the rancher's privately owned values were four dollars, the net residual federal share would be two dollars.

Purchase

Once a value has been set for the private grazing rights on a given allotment and once the residual federal share has been determined, the rancher should be given the option to purchase any residual value held by the federal government.

This approach obviously involves the forfeiture by the rancher of any right in the land except the surface rights and water rights. It does not include mineral rights or any lawful property right interests which may have developed on these lands before or since 1891. This is a major departure from a claim for full fee simple rights to the real estate which was the objective of the preemption efforts by stockmen prior to 1891. What I am proposing is a solution which will involve the greatest degree of fairness to all parties within the parameters of a workable compromise.

The stockman would obtain title to the forage subject to prior lawful rights. That title would preclude any interference by government land management agents or agencies in his ranch operation. He would be free to sell, trade, or otherwise use or dispose of his forage rights. He could sell or trade those rights to recreational interests, wildlife interests, wilderness interests on the basis of a willing seller/willing buyer transaction, allowing public demand to determine the price.

User Rights

These lands so conveyed to the rancher should be subject to the provisions of the 1872 Mining Law for exploration and mining purposes. Any reduction in forage values caused by mining activity would be compensable to the rancher using standard appraisal procedures. The procedure could be put forth in deed restrictions in the title conveyed to the rancher.

If a mining company determined that a one thousand acre parcel was needed to develop and operate a mine, applicable procedures would determine the value of the forage. If the one thousand acre parcel had a grazing capacity of one hundred AUMs, for example, and the fair market value of an AUM was thirty dollars, the compensable loss would be three thousand dollars.

These suggestions are, of course, intended simply to illustrate a principle. Actual cases could encompass any number of complexities depending on particular circumstances. However, if the value determination were conducted within existing lawful guidelines, it should be no different than a value determination on any parcel of private property with a split estate interest.

The same marketplace valuation and transaction procedure could be used to let the public's desire—expressed through their willingness to pay—determine the various ways in which the interests of all other parties would be satisfied. Oil and gas, geothermal, non-locatable minerals could all be addressed in terms of current federal leasing practices presently in place. Those recognized economic uses would be dealt with in the same general manner as in the example used for mining.

I have suggested parameters within which an equitable solution to the federal lands problem could be realized. The suggestion is stated in generalities and is not intended to address all ramifications of a many-faceted problem. It does, however, address both the stockman's interest in the land and the public's interest in terms of existing lawful procedures. It addresses the rancher's problem by the recognition of the rancher's private interest in the federal lands as a property right short of full ownership of the entire bundle of sticks, but a property right which is entitled to full due process of law plus compensation. This is the very least the stockman can settle for. This would eliminate the constant political interference in his business by the BLM and Forest Service. If the environmentalists know their politically motivated attacks on the rancher will be subject to an evidentiary hearing, cross examination, and possible penalties, the desire to attack the rancher would be greatly reduced.

The concerns and interest of the broad general public are equitably dealt with in this approach by creating an atmosphere where the rights and interests of hunters, fisherman, recreationists, and the broad public can define property rights. Within the framework of property rights, non-ranching interests can maximize the potential of the lands for wildlife, recreation, urban expansion, defense, or other relevant public concern. The market place, if allowed to work, will define and refine property rights in a manner reflecting the highest and best use as expressed by the general public, who in fact, constitute the market place.

The public also gains by the reduction in cost of supporting the two major land management agencies involved. Tax money now spent to maintain public management of private rights would be released. Those dollars would be available for needed public services elsewhere. The elimination of the inefficiency and resource waste, which tends to be the hallmark of public resource management, would terminate. In its place would be more production of consumer goods at a lower cost. Private management of the resource base, with the inherent essential of private responsibility, would provide the greatest motivation for long-term conservation of resources while sustaining production of consumer goods.

In the coming decade the rancher faces a public relations assault on his private rights unparalleled since John Muir and the Sierra Club were employed by eastern interests to attack the stockman in the 1890s. *Livestock Free by 93!* has become a rallying cry for the environmental movement. The anti-livestock philosophy is translated as de facto policy by the BLM and Forest Service into economic vandalism under the guise of resource protection. The anti-livestock philosophy is translated by the Sierra Club into efforts to elect environmentalist candidates to public office in the western states. The Natural Resources Defense Council continues to wage war on the rancher, attempting to litigate increased production costs "high enough to remove livestock from the range." Earth First! translates the anti-livestock philosophy into arson and terrorism.

The agencies and the environmentalist groups are independent entities as far as public perception is concerned. However, the arson and terrorism of Earth First! is nothing more than a logical extension of the anti-livestock philosophy of the BLM and Forest Service translated into physical violence. The BLM,

Forest Service, Sierra Club, or NRDC would all disclaim any advocacy of destruction of private property as a means to accomplish their objectives. However, none of these groups have ever used their substantial public influence to inform the public about the rancher's legitimate status as a co-owner of the split estate. It is this public ignorance that creates the atmosphere where violence can take root.

It was overkill by John Muir and the Sierra Club in their attack on livestock grazing in the 1890s that almost brought about the repeal of the Forest Reserve Act six years after its inception. The strident and sometimes violent attack on livestock interests in the 1980s, using outright lies and distortions to economically damage or destroy the livestock industry on the federal land, will generate a counter response. This time the confrontation must end in recognition of the range right as a property right, entitled to full due process of law and compensation. Anything short of that recognition must set the stage for an unacceptable level of long term political, legal, and economic turmoil which the nation can ill afford.

Two questions must be firmly resolved: (1) Exactly what is the legal status of the rancher's pre-existing property rights in the federal split estate lands; and (2) To what degree are those rights entitled to protection under the Fifth Amendment of the Constitution.

Having a strong legal position is of no value if a person fails to defend that position. Having laws on the books do not prevent crime if the individual or society doesn't demand their enforcement. A law against robbery will not stop the robber if a complaint is not filed and the penalty imposed.

For years the western rancher has relied on the political power of western congressmen. But rural power receded dramatically with the Warren court's one man, one vote ruling in *Reynolds* v. *Sims* in 1963. The rocky marriage that has existed between the land management agencies and the leadership of the livestock organizations has reached a stage of estrangement. The agencies have been wooed away from the livestock industry by the attraction of increased budgets and salaries which the environmental movement can produce through its influence on an urban Congress.

The western rancher must stand on his own. No longer can he benefit from an agreement which trades range privileges for silence on his rights.

In my opinion the western range livestock industry can challenge federal agency policy on the taking issue involving prior rights. This must be based on a well structured case using the best legal counsel. The taking issue raises the question: Has de facto federal agency policy toward the stockman been an intentional systematic denial of due process of law and the civil rights of western citizens? The ranching community of the western states must start asking the big questions.

A second approach which I believe could help resolve the rancher's current dilemma involves the revokable permit granted by the government. It would be a challenge of agency action on the premise of third party interference. The courts have repeatedly affirmed that the Forest Service and BLM have an obligation to protect the economic integrity of an unrevoked grazing permit. The courts have been consistent in affirming that the agency or its agents have no more right to interfere in the use of an unrevoked grazing permit than does any other third party.

The economic distress to livestock operators caused by such agency involvements as wild horse and burro protection on the part of wild horse groups, introduction of non-indigenous species in competition with established livestock grazing on behalf of wildlife interests, reduction of or elimination of livestock use to accommodate recreation interests in riparian areas, and many more may be grounds for an effective challenge on the basis of third party interference.

Since the shotgun marriage of the livestock associations and the Forest Service in 1905 as a way to provide special privileges for some grazing interests while suppressing the preemptive rights issue, the range livestock industry has relied on various branches of the U.S. Department of Agriculture for much of its advice and information. Current events would indicate that such advice has largely been a broken crutch. If the livestock industry is going to survive on the western range, it must trade its reliance on the local extension agent for the advice of competent legal counsel. It is true that USDA advice has been free, but the record would indicate that the industry got what it paid for.

Change is coming to the western range. The storm over rangelands, raging now for more than a hundred years, is reaching a climax. The range livestock industry may well stand or fall on its ability to mount an effective legal challenge over the issue that started it all: preemptive range rights.

Epilogue Footnotes

1. Bensel, *Sectionalism and American Political Development: 1880-1980*, p. 303.
2. Susan R. Schrepfer, "Conflict in Preservation: The Sierra Club, Save-the-Red-woods League, and Redwood National Park," *Journal of Forest History*, vol. 24, No. 2, April 1980. pp. 60-77.
3. Schrepfer here cites Joseph LeConte, *Evolution and Its Relation to Religious Thought*, D. Appleton and Company, New York, p. 325.
4. Schrepfer's footnote: For John Muir's faith in a universe of "law, order, creative intelligence, loving design," see Linnie Marsh Wolfe, *Son of the Wilderness: The Life of John Muir* (New York: Alfred Knopf, 1951), p. 77; see also William F. Badè, *The Life and Letters of John Muir* (New York: Houghton Mifflin, 1924); Duncan McDuffie Papers, Save-the-Redwoods League Office, San Francisco (hereinafter cited as SRL); David Starr Jordan and Vernon Kellogg, *Evolution and Animal Life* (New York: D. Appleton and Company, 1908), p. 563; Vernon Kellogg, *Darwinism To-day* (New York: Henry Holt, 1908), pp. 1, 56, 77.
5. Schrepfer here cites: Henry Fairfield Osborn, *Evolution and Religion in Education: The Fundamentalists Controversy of 1922-1925* (New York: Charles Scribner and Sons, 1926); Charles M. Goethe, *Garden Philosopher* (Sacramento: Keystone Press, 1955), pp. 213-15, 259, 309; John Campbell Merriam, *The Living Past* (New York: Charles Scribner and Sons, 1930).
6. Schrepfer here cites: LeConte, *Evolution and Its Relation to Religious Thought*, pp. 38-39; Osborn, *Impressions of Great Naturalists* (New York: Charles Scribner and Sons, 1928), pp. 163, 188-97; Ralph W. Chaney, "Synopsis of Lectures in Paleontology 10, University of California, Berkeley: Outline and General Principles of the History of Life," 1947, p. 80, Bancroft Library, Berkeley: Merriam, *The Living Past*, p. 137.
7. Loren Eiseley, *The Invisible Pyramid. A Naturalist Analyzes the Rocket Century* (New York: Charles Scribner's Sons, 1970), pp. 1, 40.
8. Schrepfer here cites: Ibid., pp. 16-17; Eiseley, *The Unexpected Universe* (New York: Harcourt, Brace and World, 1969), pp. 31-33.
9. Schrepfer here cites: Eiseley, *The Invisible Pyramid*, p. 35; Eiseley, *The Unexpected Universe*, p. 43.
10. Schrepfer here cites: [David] Brower, ed. *Wilderness, America's Living Heritage* (San Francisco: Sierra Club, 1962), p. 110.
11. Schrepfer here cites: Eiseley, *The Invisible Pyramid*, pp. 41, 55; Alfred G. Etter, "Jewels, Gold, and God," *Sierra Club Bulletin* 50 (September 1965): 9-10.
12. *Sierra*, published by the Sierra Club, 530 Bush Street, San Francisco, March 1989.
13. James T. Bennett and Thomas J. DiLorenzo, *Destroying Democracy: How Government Funds Partisan Politics*, Cato Institute, Washington, D. C., 1985.
14. "Who Benefits From Welfare Ranching?" by the editors of *National Boycott News*, Seattle, Spring-Summer, 1989, pp. 94-98.
15. Alan M. Gottlieb, ed., *The Wise Use Agenda*, Free Enterprise Press, Bellevue, Washington, 1989.
16. Bensel, *Sectionalism and American Political Development, 1880-1980*, pp. 303-304, 410.
17. Dallas economist John Baden is a chief proponent of the market approach to federal lands problems.
18. Opinion of Solicitor Leo M. Krulitz, M-36914, June 25, 1979, *Federal Water Rights of the National Park Service, Fish and Wildlife Service, Bureau of Reclamation, and the Bureau of Land Management*, 86 Interior Decisions 553, 571 (1979).

19. Public Water Reserve Order No. 107, *Executive Order*, April 17, 1926.

20. Opinion of Solicitor William Coldiron, September 11, 1981, *Non-Reserved Water Rights—United States Compliance with State Law*, 88 I.D. 1055, (1981).

21. C. R. Anderson and John A. VanWagenen, *A Report to the Committee on Appropriations U.S. House of Representatives on the Water Rights Policy of the Bureau of Land Management Relating to the Grazing Management Program*, Surveys and Investigations Staff, House Committee on Appropriations, November 15, 1984. The Report is marked "NOT FOR RELEASE UNTIL AUTHORIZED BY THE CHAIRMAN." The Chairman had not released the report as this book went to press in 1989. Unauthorized copies of the secret report are available from the Free Enterprise Press for cost of copying and postage.

22. 48 Stat. 1269, "Act of June 28, 1934," codified at 43 U. S. C. 315 *et seq.* Commonly known as the Taylor Grazing Act.

23. Interview with David Mathis of the Nevada Department of Wildlife, September 1989.

24. Bancroft, *History of Nevada, Colorado, and Wyoming, 1540-1888*, pp. 163-75.

25. The "Map of the State of Nevada to Accompany the Annual Report of the Commissioner of the General Land Office, 1866" is reproduced in David Beatty and Robert O. Beatty, *Nevada, Land of Discovery*, First National Bank of Nevada, Reno, 1976, p. 20.

26. Rowley, *U.S. Forest Service Grazing and Rangelands: A History*, p. 166.

27. Interview with David Mathis, September, 1989. Mathis estimated the 1955 populations from reports compiled in *Nevada Wildlife*, Nevada Fish and Game Commission Centennial Issue, 1965, edited by the Information-Education Division. In 1965 David Mathis was assistant chief of the Nevada Fish and Game Commission and information-education officer.

28. For a compilation of annual estimates from 1913 to the present, see U.S. Department of Agriculture, *Nevada—Estimated Number and Value of Livestock on Farms and Ranches—January 1*, C.E. 1-10 (3-1-63), Statistical Reporting Service, Reno, 1989, p. 1.

29. *Privatization: Toward More Effective Government*, Report of the President's Commission on Privatization, David F. Linowes, Chairman, Washington, D.C., March 1988, pp. 242-243.

TABLE OF CASES

BIBLIOGRAPHY

INDEX

TABLE OF CASES

BIBLIOGRAPHY

Unpublished Sources

Manuscripts and Manuscript Collections

Annual Grazing Report, 1912, Sopris National Forest, Sec. 63, Region 2, Dr. 31, RG 95, National Archives.

Annual Grazing Report, 1916, Rio Grande National Forest, Sec. 63, Region 2, Dr. 35, RG 95, National Archives.

G. S. Arnold to T. E. Will, June 24, 1907, Forest Service Law Office Correspondence, Record Group 49, Drawer 16, National Archives.

Richard A. Bartlett, *The Genesis of National Park Concessions Policy*, Colloquium Paper 1980, Restricted, National Park Service Archives, Harpers Ferry, West Virginia.

"Description of Township 16 S., R 31 E, W M in Grant County, Oregon," RG 95, Entry 13 (43), Vol H, National Archives.

"Description of Township 10 South, Range 33 E. W. M. in Grant County, Oregon," RG 95, Entry 13 (45), Vol H, National Archives.

"Description of Township 15 South, Range 26 East, W. M. in Grant County, Oregon," RG 95, Entry 13, (42) Vol H, National Archives.

Bernhard Fernow to H. G. deLotbiniere, March 6, 1897, RG 95-2, National Archives.

Hague Papers, letter book 3B, R. G. 57, National Archives. Hague to E. M. Dawson, March 16, Hague to Brown, March 16, 1891, Hague to Noble, March 25, 1891, Hague to Grinnell, April 4, 1891, Hague to Grinnell, April 6, 1891.

Russell B. Harrison to Benajmin Harrison, October 24, 1885. Wyoming Stock Growers Association Collection, Western History Research Center, University of Wyoming.

Archer Butler Hulbert, ed., "The Records of the Original Proceedings of the Ohio Company," *Marietta College Historical Collections*, I and II.

William Jones, Sworn statement by William Jones, rancher, Eddy County, New Mexico, April 10, 1917. R. G. 49, Unlawful Enclosure, Box 799, National Archives.

"The Legal Aspects of the Grazing Problem, by Ethelbert Ward, Denver, Colorado," Forest Service Law Office Correspondence RG 49, Drawer 16, National Archives.

E. H. Libby to James Glendenning, U. S. Forestry Supintendent, Lewiston, Idaho. Forest Service Field Records, 1898-1904. R. G. 95 National Archives.

Duncan McDuffie Papers, Save-the-Redwoods League Office, San Francisco.

Memorandum Opinion, No. 725, Circuit Court of the United States, Ninth Circuit, District of Montana, in Forest Service Law Office Correspondence RG 49, Drawer 16, National Archives.

W. J. Palmer to Pinchot, May 23, 1907, Forest Service Law Office Correspondence, R. G. 49, Drawer 16, National Archives.

W. J. Palmer to P. P. Wells, (No date), Forest Service Law Office Correspondence, R. G. 49, Drawer 16, National Archives.

People of Grant County, Oregon, *Protest and Remonstrance: The People of Grant County, Oregon, Object to More Than Half of the County's Area Being Withdrawn from Settlement,* Forest Service Field Records, 1898-1904. R. G. 95 National Archives.

"Protest Against Forest Reserve," RG 95, Entry 13 (12), Vol. H, National Archives.

J. A. E. Psujdam to Binger Hermann, December 14, 1899. Forest Service Field Records, 1898-1904, R. G. 95 National Archives.

Pueblo Chieftain, May 15, 1907. Forest Service Law Office Correspondence, R. G. 49, Drawer 16, National Archives.

"To the editor" [of an unidentified Oregon newspaper], RG 95, Entry 13 (47), Vol H, National Archives.

1956 Rangers Training Camp, "Range Administrators," May 7, 1956, Box 3, 13/2/1:3, R. G. 95, Federal Records Center for the National Archives, Denver, Colo.

William Sparks, General Land Office Commissioner, to Interior Secretary Lucius Lamar, August 19, 1885. R. G. 48, Office of the Secretary of the Interior, Central Classified files, National Archives.

"[Washington State] House Memorial No. 6 [to the President, his Cabinet, and Congress]," signed by J. A. Falconer, Speaker of the House, and Charles E. Coon, President of the Senate, RG 95, Entry 13 (7), Vol H, National Archives.

P. P. Wells to T. E. Will, June 29, 1907, Forest Service Law Office Correspondence, R. G. 49, Drawer 16, National Archives.

Theses and Carnets

Ralph W. Chaney, "Synopsis of Lectures in Paleontology 10, University of California, Berkeley: Outline and General Principles of the History of Life," Bancroft Library, Berkeley, 1947.

Herbert D. Kirkland, *The American Forests, 1864-1898: A Trend Toward Conservation,* unpublished Ph.D. dissertation, University of Florida, 1971.

Interviews

David Mathis, Nevada Department of Wildlife, September 1989.

Dr. Harold K. Steen, executive director of the Forest History Society, telephone interview with editor Ron Arnold, September, 1988.

Published Sources

Government Publications and Documents

American State Papers: Documents, Legislative and Executive, of the Congress of the United States, 1789-1838. 38 vols., arranged by subject matter.

Washington: Gales & Seaton, 1832-1861.

C. R. Anderson and John A. VanWagenen, *A Report to the Committee on Appropriations U.S. House of Representatives on the Water Rights Policy of the Bureau of Land Management Relating to the Grazing Management Program*, Surveys and Investigations Staff, House Committee on Appropriations, November 15, 1984.

"Attorney General Guidelines for the Evaluation of Risk and Avoidance of Unanticipated Takings." (Justice Department)

Fred Bosselman, David Callies and John Banta, *The Taking Issue: An Analysis of the Constitutional Limits of Land Use Control*, Council on Environmental Quality, Washington, D. C., 1973.

Congressional Globe, 27th Congress, 1st Session.

Congressional Record, House, March 27, 1888, 50 Cong. Sess. 1, p. 2463.

Congressional Record, House, June 26, 1888, 50 Cong. Sess. 1, 1888, pp. 5592-5593.

Congressional Record, House, 51st Cong., 1st Sess., June 25, 1890.

Congressional Record, Senate, 51st Cong., 2d sess., February 28, 1891, p. 3547.

Congressional Record, House, 51st Cong., 2d sess., March 2, 1891, pp. 3611-16.

Congressional Record, 53rd Cong., 1 Sess., Vol. XXV, Pt. 2, pp. 2434-35.

Congressional Research Service, *The Major Federal Land Management Agencies: Management of our Nation's Lands and Resources,* Document 87-22 ENR, Library of Congress, Washington, January 14, 1987.

Frederick V. Coville, *Forest Growth and Sheep Grazing in the Cascade Mountains of Oregon*, U. S. Department of Agriculture, Division of Forestry Bulletin No. 15.

Disposal of Public Lands, House of Representatives Report No. 778, 50th Congress, 1st Sess., Feb. 28, 1880, Washington.

Thomas Donaldson, *The Public Lands*, Government Printing Office, Washington, D.C., 1884.

Environmental Law Institute, *The Evolution of National Wildlife Law*, prepared for the Council on Environmental Quality, Government Printing Office, Washington, 1977.

Executive Order, Public Water Reserve Order No. 107, April 17, 1926.

Executive Order No. 12630, *Governmental Actions and Interference With Constitutionally Protected Property Rights*, Federal Register, vol. 53, No. 53, p. 8859, Friday, March 18, 1988.

Bernhard Eduard Fernow, *Report upon the Forestry Investigation of the United States Department of Agriculture, 1877-1898*, Government Printing Office, Washington, D. C., 1899.

Forest Service, U. S. Department of Agriculture, *The Use of the National Forest Reserves,* Washington, D. C., July, 1905.

Paul W. Gates and Robert W. Swenson, *History of Public Land Law Development*, Public Land Law Review Commission, Washington, D.C., 1968.

House Committee on Public Lands, *Hearings on grazing, homesteads, and the regulation of grazing on the public lands,* 1914.

Wells A. Hutchins, *Water Rights Laws in the Nineteen Western States*, Miscellaneous Publication No. 1206, Natural Resource Economic Division, Economic Research Service, United States Department of Agriculture, Washington, D.C., 1971.

Testimony of Secretary of the Interior Harold L. Ickes, Taylor Grazing Act Senate Hearings, *Legislative History*.

Interior Board of Land Appeals, Case No. 84-88, May 22, 1984.

Worthington Chauncey Ford et al., editors, *Journals of the Continental Congress*, 34 vols., Government Printing Office, Washington, Washington, 1904-1937.

Land Office Report, Government Printing Office, Washington, 1888.

Laws and Ordinances of the State of Deseret, Compilation 1851, Salt Lake City, 1919.

Laws of the United States, Vol. X.

"Maryland's instructions to her delegates in Congress," Dec. 15, 1778, in *Johns Hopkins University Studies*, III, No. 1.

One Third of the Nation's Land: A Report to the President and to the Congress by the Public Land Law Review Commission, Washington, D. C., June 1970.

Opinion of Solicitor Leo M. Krulitz, M-36914, June 25, 1979, *Federal Water Rights of the National Park Service, Fish and Wildlife Service, Bureau of Reclamation, and the Bureau of Land Management*, 86 Interior Decisions 553 (1979).

Opinion of Solicitor William Coldiron, September 11, 1981, *Non-Reserved Water Rights—United States Compliance with State Law*, 88 I.D. 1055, (1981).

John W. Powell, *Report on the Lands of the Arid Region of the United States*, 45th Cong. 2d sess., 1878, H. Exec. Doc. 73.

Privatization: Toward More Effective Government, Report of the President's Commission on Privatization, David F. Linowes, Chairman, Washington, D.C., March 1988.

Report and Recommendations of Committee of National Reclamation Association, "Preservation of Integrity of State Water Laws," (1943).

Report of the Commissioner of the General Land Office on the Unauthorized Fencing of the Public Lands, *Sen. Ex. Doc.* No. 127, 48 Cong., Sess. 1, 1883-1884.

Report of the Secretary of the Interior, 1887.

James D. Richardson, *A Compilation of the Messages and Papers of the Presidents, 1789-1897*, Government Printing Office, Washington, 1896-1899, Vol. 1.

H. M. Taylor, "Importance of the Range Cattle Industry," *Annual Report* of the Bureau of Animal Husbandry, Washington, 1886.

Treaty of Paris of 1783, *Treaties and Conventions Concluded Between the United States of America and Other Powers Since July 4th, 1776*, Government Printing Office, Washington, 1889.

U. S. Congress, House, *House Resolution 7901*, 50th Cong., 1st Sess., 1888, introduced by William S. Holman February 29, 1888.

U. S. Congress, House, 50th Cong., 1st Sess., 1888, *Report No. 778*, signed February 25, 1888 by William S. Holman.

U. S. Congress, House, *Public Timber Lands*, 50th Cong., 1st Sess., 1888, H. Ex. Doc. 242.

U. S. Congress, *Senate Report of the Committee Appointed by the National Academy of Sciences upon the Inauguration of a Forest Policy for the Forested Lands of the United States*, 55th Cong., 1st Sess., 1897, S. Doc. 105, vol. 5.

U. S. Congress, Senate, *A Report on the Western Range: A Great But Neglected Natural Resource*, 74th Cong., 2d Sess., 1936, S. Doc. 199.

U.S. Department of Agriculture, *National Plan for American Forestry*, I, Washington, D. C., 1933.

U.S. Forest Service, *The Use of the National Forest Reserves,* July, 1905.

The United States Government Manual, National Archives and Records Service, Washington, D.C., 1988.

Newspapers

Cheyenne, Wyoming, *Leader*, May 8, 1868.

Kaskaskia (Illinois) *Western Intelligencer*, Nathaniel Pope article, January 21, 1818.

New York Times, September 18, 1885, August 5, 1888, December 21, 1891.

USA Today, Editorial Faceoff, May 3, 1989.

Articles

George F. Authier, "Both Sides of the Range Controversy," *American Forests and Forest Life* 31 (December, 1925).

Walter Dean Burnham, "The End of American Party Politics" (Transaction) *Society*. vol. 7, No. 2, December 1969.

J. Baird Callicott, "American Indian Land Wisdom? Sorting Out the Issues," *Journal of Forest History*, Vol. 33, No. 1, January 1989.

William E. Colby, "The Freedom of the Miner and It Influence on Water Law," published in *Legal Essays, in Tribute to Orrin Kipp McMurray*, 1935.

R. Taylor Dennen, "Cattlemen's Associations and Property Rights in Land in the American West," *Explorations in Economic History*, vol. 13.

Donahue, "The Future of the Concept of Property Predicted from the its Past," in XXII *Nomos: Property*.

Alfred G. Etter, "Jewels, Gold, and God," *Sierra Club Bulletin* 50 (September 1965).

Forest History, "Philip P. Wells in the Forest Service Law Office," April, 1972.

"The Forestry Meeting at Lancaster," *Forest Leaves*, III (April, 1890).

Rudolf Freund, "Military Bounty Lands and The Origins of the Public Domain," in *The Public Lands: Studies in the History of the Public Domain*.

Eric T. Freyfogle, "Land Use and the Study of Early American History," *Yale Law Journal*, vol. 94, 1985.

William B. Greeley, "The Stockmen and the National Forests," *Saturday Evening Post*, November 14, 1925.

Gene M. Gressley, "The American Cattle Trust: A Study in Protest," *Pacific Historical Review*, February 1961.

Garrett Hardin, "The Tragedy of the Commons," *Science*, December 13, 1968.

A. M. Honoré, "Ownership," in *Oxford Essays in Jurisprudence*, ed. A. G. Guest, Oxford University Press, Oxford, 1961.

Wells A. Hutchins, "The Community Acequia: Its Origin and Development," Southwestern Historical Quarterly, vol. 31, 1928.

The Adverse Economic Impact of Wilderness Land Withdrawals in Elko County, Nevada, George F. Leaming, Ph.D., Western Economic Analysis Center. Presented at the Second Annual National Wilderness Conference, Reno, Nevada, April, 1989.

Thomas M. Lenard, "Wasting Our National Forests—How to Get Less Timber and Less Wilderness at the Same Time," *Regulation*, July/August 1981.

Ray H. Mattison, "Roosevelt and the Stockmen's Association," *North Dakota History*, XVII (April 1950).

Charles W. McCurdy, "Stephen J. Field and Public Land Law Development in California, 1850-1866: A Case Study of Judicial Resource Allocation in Nineteenth-Century America," *Law and Society* (Winter) 1976.

Robert H. Nelson, "Why the Privatization Movement Failed," *Regulation*, July/August 1984.

Robert H. Nelson, "Why the Sagebrush Revolution Burned Out," *Regulation*, May/June 1984.

Albert F. Potter, "Cooperation in Range Management," *American National Cattleman's Association Proceedings*, 16 (1913).

Proceedings of the American Forestry Association, XII.

Public Lands News, Summer, 1988.

Sierra, published by the Sierra Club, 530 Bush Street, San Francisco, March 1989.

Susan R. Schrepfer, "Conflict in Preservation: The Sierra Club, Save-the-Redwoods League, and Redwood National Park," *Journal of Forest History*, vol. 24, No. 2, April 1980.

Charles H. Shinn, "Work in a National Forest," *Forestry and Irrigation*, vol. 13, November, 1907.

Payson Jackson Treat, "Origin of the National Land System under the Confederation," in *The Public Lands: Studies in the History of the Public Domain*, edited by Vernon Carstensen, University of Wisconsin Press, Madison, 1962.

Published Correspondence

John Adams to Alexander Everett, May 24, 1830, and September 18, 1831, *American Historical Review*, 11, January, 1906.

James Campbell to Allen Trimble, February 11, 1819, in *Old Northwest Genealogical Quarterly*, vol. 10, July, 1907.

Rufus King to Christopher Gore, July 17, 1819, February 11, 1819, and April 9, 1820, in *The Life and Correspondence of Rufus King*, ed. Charles R. King, New York, Putnam's Sons, 1894-1900, vol. 6.

George Washington to the President of Congress, Sept. 2, 1776, in Peter Force, ed., *American Archives*, series 5, 3 vols., Washington, 1848-1853.

Books

Kenneth D. Ackerman, *The Gold Ring: Jim Fisk, Jay Gould, and Black Friday*, Dodd, Mead, New York, 1988.

John Adams, *Works*, edited by C. F. Adams, Vol IX, Little, Brown & Company, Boston, 1854.

Ramon Frederick Adams, *The Rampaging Herd*, University of Oklahoma Press, Norman, 1959.

Edward Jones Allen, *United Order Among the Mormons*, Studies in History, Economics and Public Science #419, Columbia University, New York, 1936.

Aristotle, *Politics*, translated by H. Rackham, Harvard University Press, Cambridge.

Ron Arnold, *Ecology Wars: Environmentalism as if People Mattered*, Free Enterprise Press, Bellevue, 1987.

Leonard Arrington, *The Changing Economic Structure of the Mountain West, 1850-1950*, Utah State University Press, Logan, 1963.

William F. Badè, *The Life and Letters of John Muir*, Houghton Mifflin, New York, 1924.

Hubert Howe Bancroft, *History of Nevada, Colorado, and Wyoming, 1540-1888*, The History Company, San Francisco, 1890.

Hubert Howe Bancroft, *History of Utah, 1540-1887*, The History Company, San Francisco, 1890.

Will Croft Barnes, *Apaches and Longhorns: The Reminiscences of Will C. Barnes*, University of Arizona Press, Tucson, 1982.

James T. Bennett and Thomas J. DiLorenzo, *Destroying Democracy: How Government Funds Partisan Politics*, Cato Institute, Washington, D. C., 1985.

Richard Franklin Bensel, *Sectionalism and American Political Development: 1880-1980*, The University of Wisconsin Press, Madison, 1984.

Ralph Paul Bieber, ed., *Southern Trails to California in 1849*, The Arthur H. Clark Company, Glendale, 1937.

Israel George Blake, *The Holmans of Veraestau*, The Mississippi Valley Press, Oxford, Ohio, 1943.

Black's Law Dictionary, Fifth Edition, West Publishing Company, St. Paul, 1979.

Sir William Blackstone, *Commentaries on the Laws of England*, 4 vol., 12th edition, edited by Edward Christian, Cambridge, London, 1794.

Jay Boyd, *The Crédit Mobilier of America*, C. W. Calkins, Boston, 1880, reprinted 1969.

Charles Bray, *Financing the Western Cattle Industry*, Fort Collins, 1928.

Brower, ed., *Wilderness, America's Living Heritage*, Sierra Club, San Francisco, 1962.

George Rothwell Brown, ed., *Reminiscences of William M. Stewart of Nevada*, Neale Publishing Company, New York, 1908.

Sterling Brubaker, ed., *Rethinking the Federal Lands*, Resources for the Future, Washington, D.C., 1984.

Edmund Burke, "Speech on Conciliation with America," 1775, *The Works of Edmund Burke*, Little, Brown and Company, Boston, 1899.

Jenks Cameron, *The Development of Government Forest Control in the United States*, Johns Hopkins University Press, Baltimore, 1928.

Robert A. Caro, *The Years of Lyndon Johnson: The Path to Power*, Random House, New York, 1981.

Cicero, *De Officiis*, translated by Walter Miller, Loeb edition, Harvard University Press, Cambridge.

Cochran's Law Lexicon, Fourth Edition, The W. H. Anderson Company, Cincinnati, 1956.

Corporate Foundation Profiles, The Foundation Center, eds., New York, 1988.

William Augustus Croffut, *The Vanderbilts and the Story of Their Fortune*, Belford, Clarke, Chicago, 1886, reprinted by Arno Press, New York, 1975.

Hezekiah John Crumpton, *The Adventures of Two Alabama Boys*, (self-published), Montgomery, 1912.

Alfred G. Cuzán, "Appropriators Versus Expropriators: The Political Economy of Water in the West," in *Water Rights: Scarce Resource Allocation, Bureaucracy, and the Environment*, Pacific Institute for Public Policy Research, San Francisco, 1983.

C. J. Czajkowski, *The Theory of Private Property in Locke's Political Philosophy*, University of Notre Dame Press, South Bend, 1941.

Samuel T. Dana and Sally K. Fairfax, *Forest and Range Policy: Its Development in the United States.*, 2d ed., McGraw-Hill, New York.

Richard C. Davis, *North American Forest History: A Guide to Archives and Manuscripts in the United States and Canada*, ABC-Clio Press, Santa Barbara, 1976.

Leslie Decker, *Railroads, Land and Politics*, Brown University Press, Providence, 1964.

The Journals of Alfred Doten, 1849-1903, Walter Van Tilberg Clark, ed., vol. 1, University of Nevada Press, Reno, 1973.

Isaac H. Duval, *Texas Argonauts: Isaac H. Duval and the California Gold Rush*, Book Club of California, San Francisco, 1987.

Echo Bay Mines, *Annual Report 1987*, Denver.

Loren Eiseley, *The Invisible Pyramid, A Naturalist Analyzes the Rocket Century*, Charles Scribner's Sons, New York, 1970.

Loren Eiseley, *The Unexpected Universe*, Harcourt, Brace and World, New York, 1969.

Chris Emmett, *Shanghai Pierce*, University of Oklahoma Press, Norman, 1953.

Environmental Law, Protection and Education, The Foundation Center, Comsearch Grant Guides series #22, New York, 1989.

Richard A. Epstein, *Takings: Private Property and the Power of Eminent Domain*, Harvard University Press, Cambridge, 1985.

P. N. Fedoseyev, Irene Bakh, L. I. Golman, N. Y. Kolpinsky, B. A. Krylov, I. I. Kuzminov, A. I. Malysh, V. G. Mosolov, Yevgenia Stepanova, Institute of Marxism-Leninism of the Communist Party of the Soviet Union, *Karl Marx: A Biography,* Progress Publishers, Moscow, 1973.

Ronald H. Fahl, *North American Forest and Conservation History: A Bibliography*, ABC-Clio Press, Santa Barbara, 1976.

Daniel Feller, *The Public Lands in Jacksonian Politics*, University of Wisconsin Press, Madison, 1984.

Bernhard Eduard Fernow, *Economics of Forestry*, Thomas Y. Crowell and Company, New York, 1902.

The Foundation Grants Index, The Foundation Center, New York, 1988.

John G. Francis and Richard Ganzel, eds., *Western Public Lands: The Management of Natural Resources in a Time of Declining Federalism*, Rowman & Allanheld, Totowa, New Jersey, 1984.

Bernard J. Frieden, *The Environmental Protection Hustle*, The MIT Press, Cambridge, 1979.

Ray Ginger, *Age of Excess: The United States from 1877 to 1914*, Macmillan, New York, 1965.

Charles M. Goethe, *Garden Philosopher*, Keystone Press, Sacramento, 1955.

Gene M. Gressley, *Bankers and Cattlemen*, Alfred A. Knopf, New York, 1966.

Samuel P. Hays, *Conservation and the Gospel of Efficiency: The Progressive Conservation Movement 1890-1920*, Harvard University Press, Cambridge, 1959.

William Waller Hening, *The Statutes at Large; Being a Collection of All the Laws of Virginia, 1619-1792* (13 vols., Richmond, 1809-1823), vol. 10.

Franz Herderhold, ed., *Das Urbar, Der Grafshaft Ravenberg, Von 1556*, [The Cadastral Register of the Earldom of Ravenberg, from 1556], Bearbeitet Von Franz Herderhold, Band I Tex, Aschendorffsche Verlagsbuchhandlung, Munster, 1960.

Benjamin Horace Hibbard, *A History of the Public Land Policies*, Macmillan Company, New York, 1924, reprinted by the University of Wisconsin Press, Madison, 1965.

History of Nevada, Thompson & West, Oakland, 1881, reprinted by Howell-North, Berkeley, 1958.

Stewart H. Holbrook, *The Age of the Moguls*, Doubleday & Company, Inc., Garden City, New York, 1956.

Don Hummel, *Stealing the National Parks: The Destruction of Concessions and Park Access*, Free Enterprise Press, Bellevue, 1987.

John Ise, *The United States Forest Policy*, Yale University Press, New Haven, 1920.

Robert Underwood Johnson, *Remembered Yesterdays*, Little, Brown and Company, Boston, 1923.

David Starr Jordan and Vernon Kellogg, *Evolution and Animal Life*, D. Appleton and Company, New York, 1908.

Matthew Josephson, *The Robber Barons: the Great American Capitalists*, 1861-1901, Harcourt, Brace, New York, 1934.

Vernon Kellogg, *Darwinism To-day*, Henry Holt, New York, 1908.

Maury Klein, *The Life and Legend of Jay Gould*, Johns Hopkins University Press, Baltimore, 1986.

Richard D. Lamm and Michael McCarthy, *The Angry West: A Vulnerable Land and Its Future*, Houghton Mifflin Company, Boston, 1982.

Nathaniel Pitt Langford, *The Discovery of Yellowstone Park: Journal of the Washburn Expedition to the Yellowstone and Firehole Rivers in the Year 1870*, University of Nebraska Press, Lincoln, 1972.

Henrietta Larson, *Jay Cooke*, Harvard University Press, Cambridge, 1936.

Joseph LeConte, *Evolution and Its Relation to Religious Thought*, D. Appleton and Company, New York, 1889.

Gary Libecap, *Locking Up the Range: Federal Land Controls and Grazing*, Pacific Institute for Public Policy Research, San Francisco, 1981.

Thomas L. Livermore, *Numbers and Losses in the Civil War, 1861-1865*, Houghton Mifflin & Co., Boston, 1900, reprint Morningside, Dayton, 1986.

John Locke, *First Treatise of Government*, edition of Peter Laslett, Cambridge University Press, Cambridge, 1960.

John Locke, *The Second Treatise of Government*, edition of Thomas P. Peardon, Macmillan Publishing Company, New York, 1985.

G. Michael McCarthy, *Hour of Trial: The Conservation Conflict in Colorado and the West, 1891-1907*, University of Oklahoma Press, Norman, 1977.

James M. McPherson, *Ordeal By Fire: The Civil War and Reconstruction*, Alfred A. Knopf, New York, 1982.

Tibor R. Machan and M. Bruce Johnson, eds., *Rights and Regulation: Ethical, Political, and Economic Issues*, Pacific Institute for Public Policy Research, San Francisco, 1983.

Effie Mona Mack, *Nevada*, Arthur H. Clark Company, Glendale, California, 1935.

Karl Marx, *Capital: A Critique of Political Economy, Volume 1, The Process of Capitalist Production* (1867), translated by Samuel Moore and Edward Aveling, edited by Frederick Engels, International Publishers, New York, 1967.

Ray H. Mattison, *Roosevelt's Dakota Ranches*, Bismark Tribune, Bismark, North Dakota, 1958.

Elwood Mead, *Irrigation Institutions*, Macmillan Company, New York, 1903.

John Campbell Merriam, *The Living Past*, Charles Scribner and Sons, New York, 1930.

Charles-Louis de Secondat, Seigneur de Montesquieu, *The Spirit of Laws*, translated by Thomas Nugent, Nourse, London, 1750.

John Muir, "A Windstorm in the Forests," *Mountains of California*, vol. I, Houghton Mifflin, Boston, 1917.

Henry Fairfield Osborn, *Evolution and Religion in Education: The Fundamen-*

talists Controversy of 1922-1925, Charles Scribner and Sons, New York, 1926.

Henry Fairfield Osborn, *Impressions of Great Naturalists*, Charles Scribner and Sons, New York, 1928.

Ernest Staples Osgood, *The Day of the Cattleman*, The University of Chicago Press, Chicago, 1929.

Ellen Frankel Paul, *Property Rights and Eminent Domain*, Transaction Books, New Brunswick, 1987.

Richard F. Pettigrew, *Triumphant Plutocracy: The Story of American Public Life from 1870 to 1920*, The Academy Presss, New York, 1922.

Joseph M. Petulla, *American Environmental History: The Exploitation and Conservation of Natural Resources*, Boyd & Fraser Publishing Company, San Francisco, 1977.

Gifford Pinchot, *Breaking New Ground*, Harcourt, Brace, and Company, New York, 1947.

Plato, *Laws*, translated by Rev. R. G. Bury, Loeb edition, Harvard University Press, Cambridge.

Plato, *Republic*, translated by Paul Shorey, Loeb edition, Harvard University Press, Cambridge.

Karl R. Popper, *The Open Society and Its Enemies*, vol. 1, *The Spell of Plato*, Princeton University Press, Princeton, 1962.

Proceedings of the American Forestry Association, XI, 162.

The Public Statutes at Large of the United States of America, Little and Brown, Boston, 1845-1848.

Lawrence Rakestraw, *History of Forest Conservation in the Pacific Northwest 1891-1913*, New York, 1979.

J. G. Randall and David Donald, *The Civil War and Reconstruction*, 2nd ed., revised, Heath, Lexington, 1969.

Conyers Read, *Mr. Secretary Walsingham*, 1924, reprinted Knopf, New York, 1967.

John F. Reiger, *American Sportsmen and the Origins of Conservation*, Winchester Press, New York, 1975, reprinted by University of Oklahoma Press, Norman, 1986.

Andrew Denny Rodgers III, *Bernhard Eduard Fernow: A Story of North American Forestry*, Princeton University Press, Princeton, 1951.

Paul Henle Roberts, *Hoof Prints on Forest Ranges: The Early Years of National Forest Range Administration*, Naylor, San Antonio, 1963.

Malcolm J. Rohrbough, *The Land Office Business*, Oxford University Press, New York, 1968.

Jean-Jacques Rousseau, *Discourse on Political Economy*. First appeared as an article in Diderot's *Encyclopédie* (1755). See the modern English edition in *Of the Social Contract* and *Discourse on Political Economy*, translated by Charles M. Sherover, Harper & Row, New York, 1984.

William D. Rowley, *U.S. Forest Service Grazing and Rangelands: A History*, Texas A&M University Press, College Station, 1985.

Richard Schlatter, *Private Property: The History of an Idea*, Rutgers University

Press, New Brunswick, New Jersey, 1951.

William B. Scott, *In Pursuit of Happiness, American Conceptions of Prosperity from the Seventeenth Century to the Twentieth Century*, University of Indiana Press, Bloomington, 1977.

James G. Scrugham, ed., *Nevada: A Narrative of the Conquest of a Frontier Land*, 3 vol., American Historical Society, Chicago and New York, 1935.

Charles Howard Shinn, *Mining Camps, A Study in American Frontier Government*, Charles Scribner's Sons, New York, 1885, reprint Knopf, New York, 1947.

Milton H. Shutes, *Lincoln and California*, Stanford University Press, Stanford, 1943.

Alfred William Brian Simpson, *An Introduction to the History of the Land Law*, Oxford University Press, Toronto, 1961.

Adam Smith, *An Inquiry into the Nature and Causes of the Wealth of Nations*, first edition, London, 1776.

K. A. Soderberg and Jackie DuRette *People of the Tongass: Alaska Forestry Under Attack*, Free Enterprise Press, 1989.

Harold K. Steen, *The U.S. Forest Service, A History*, University of Washington Press, Seattle, 1976.

George Thomas, *Early Irrigation in the Western States*, University of Utah, Salt Lake City, 1948.

Payson Jackson Treat, *The National Land System, 1785-1820*, E. B. Treat & Co., New York, 1910.

William Tucker, *Progress and Privilege: America in the Age of Environmentalism*, Doubleday, New York, 1982.

John Robert Umbeck, *A Theory of Property Rights, With Application to the California Gold Rush*, Iowa State University, Ames, 1981.

Orville J. Victor, *The History, Civil, Political and Military, of the Southern Rebellion, From its Incipient Stages to its Close*, James D. Torrey, New York, 1861.

William Walker, *The War in Nicaragua*, S. H. Goetzel, Mobile, 1860; reprinted B. Ethridge Books, Detroit, 1971.

Eugene F. Ware, *Roman Water Law, translated from The Pandects of Justinian, containing all of the Roman law, concerning fresh water, found in the Corpus Juris Civilis.* West Publishing Company, St. Paul, 1905.

E. H. Warmington, translator, *Remains of Old Latin*, vol III, *Lucilius* and *Laws of the XII Tables*, Loeb edition, Harvard University Press, Cambridge, 1938.

Walter Prescott Webb, *The Great Frontier*, Houghton Mifflin, Boston, 1952.

Walter Prescott Webb, *The Great Plains*, Ginn & Co., Boston, 1931.

Samuel C. Wiel, *Water Rights in the Western States*, Bancroft-Whitney Company, San Francisco, 1980.

Linnie Marsh Wolfe, *Son of the Wilderness: The Life of John Muir*, The University of Wisconsin Press, Madison, 1945.

INDEX

If you have enjoyed STORM OVER RANGELANDS, you'll want to own these other exciting titles from The Free Enterprise Press.

It Takes a Hero: The Grassroots Battle Against Environmental Oppression, by William Perry Pendley, 348 pages, paperback, $14.95.

Trashing the Economy: How Runaway Environmentalism is Wrecking America, by Ron Arnold and Alan Gottlieb. 672 pages, paperback, $19.95.

The Asbestos Racket: An Environmental Parable, by Michael J. Bennett 256 pages, paperback, $9.95.

Stealing the National Parks: The Destruction of Concessions and Public Access, by Don Hummel, 428 pages, hardcover, $19.95.

Ecology Wars: Environmentalism As If People Mattered, by Ron Arnold, 182 pages, paperback, $14.95.

The Wise Use Agenda, edited by Alan Gottlieb, 168 pages, paperback, $9.95.

Yes! Iwant to own: (please note quantity of each title ordered)
___It Takes A Hero   ___The Asbestos Racket
___Trashing the Economy   ___Ecology Wars
___The Wise Use Agenda   ___Stealing the National Parks
Total price of book(s) ordered (Prices listed above) $_____.
Add $2.00 shipping and handling for each book $_____.
Total amount (books plus S & H) this order: $_____.
Free Enterprise Press books are distributed by Merril Press.
___Check made to Merril Press ___Visa ___MasterCard
Card Number_____Expires_____
Name_____
Street_____
City_____State___ZIP_____
Mail to: Merril Press, PO Box 1682, Bellevue, WA 98009
Telephone orders: 206-454-7009 FAX orders: 206-451-3959